Introduction to CATIA V5
Release 16
(A Hands-On Tutorial Approach)

Kirstie Plantenberg
University of Detroit Mercy

ISBN: 1-58503-325-1

PUBLICATIONS

Schroff Development Corporation

www.schroff.com
www.schroff-europe.com

Examination Copies:

Electronic Files:

Trademarks

CATIA is a registered trademark of Dassault Systemes.
IBM is a registered trademark of International Business Machines.
Windows XP is a registered trademark of Microsoft Corporation.

PREFACE

An Introduction to CATIA (A Hands-On Tutorial Approach) is a collection of tutorials meant to familiarize the reader with CATIA's mechanical design workbenches. The reader is not required to have any previous CATIA knowledge. The workbenches covered in this book are; Sketcher, Part Design, Wireframe and Surface Design, Assembly Design, and Drafting. Preceding each tutorial is a brief description of the workbench, toolbars, and commands to be used and focused on within the tutorial. This book is not meant to be strictly a reference book. It is meant to enable the reader to get right into CATIA and start drawing. You will learn by doing. The author directs the reader to CATIA's Companion for in depth reference material and a more detailed description of the commands.

Changes from Release 12

For the most part, the commands and methods of constructing a part or an assembly within the *Mechanical Design* workbenches has remained the same. The biggest change between Release 12 and Release 16 is the visual look and placement of the command icons within the toolbars. Some command icons have changed appearance, others have moved to different toolbars and there were a few added commands. To help the instructor who is familiar with the Release 12 edition of *An Introduction to CATIA*, changes are indicated by a change in the toolbar color. For instance toolbars that have not changes between release 12 and 16 are dark in color. If the command icons within a toolbar have changed or a command has been added, the toolbar will be light in color.

Toolbar V12 and V16   Toolbar V16

The author has also noticed that the *Reorder* command has become highly unstable and does not work as well as it did in Release 12.

Suggested time line for a 15 week course

CATIA Tutorials	CATIA Exercises	Week
CATIA Basics: Tutorials 1.1, 1.2 & 2.1	Exercise 2.1	Week 1
Sketcher: Tutorials 2.2, 2.3	Exercise 2.2, 2.3	Week 2
Sketcher: Tutorials 2.4, 2.5 & 2.6	Exercise 2.4, 2.5, 2.6	Week 3
Part Design: Tutorials 2.7, 3.1 & 3.2	Exercise 3.1, 3.2	Week 4
Part Design: Tutorials 3.3 & 3.4	Exercise 3.3	Week 5
Part Design: Tutorial 3.5	Exercise 3.4	Week 6
Part Design: Tutorials 3.6 & 3.7	Exercise 3.5	Week 7
Part Design: Tutorials 3.8 & 3.9		Week 8
Wireframe: Tutorial 4.1		Week 9
Wireframe: Tutorials 4.2 & 4.3	Exercise 4.1	Week 10
Assembly: Tutorial 5.1		Week 11
Assembly: Tutorial 5.2	Exercise 5.1	Week 12
Drafting: Tutorial 6.1	Exercise 6.1	Week 13

NOTE: One week is left available for spring break or computer mishaps. Another week is left available for the final.

ACKNOWLEDGMENTS

I would like to thank my mother, Phyllis Plantenberg, for painstakingly going over and performing each tutorial. This was an invaluable service enabling me to correct mistakes, misinterpretations, and vagueness. I would also like to thank the University of Detroit Mercy for providing the tools necessary to make this book possible.

TABLE OF CONTENTS

Chapter 5: Assembly Design Fundamentals

Chapter 6: Drafting Fundamentals

Chapter 1: CATIA BASICS

Introduction

Chapter 1 focuses on CATIA's interface and file management. The tutorials within this chapter are meant to help the reader navigate through CATIA's workbenches as well as to show how objects are selected.

!Note to UNIX users!

The tutorials within this book were designed for the Windows version of CATIA. There may be slight differences in how CATIA looks. One operational difference is this; in the UNIX version, to multiple select objects while in a command, the user must hold the CTRL key down. If the user does not do this, the current command window will disappear behind the document window.

Tutorials Contained in Chapter 1

- Tutorial 1.1: CATIA's interface and file management
- Tutorial 1.2: Selecting, Editing & Viewing Objects

NOTES:

Chapter 1: CATIA BASICS

Tutorial 1.1:
CATIA's Interface and file management

Featured Topics & Commands

Prerequisite Knowledge & Commands

- The windows operating system

CATIA's Interface

CATIA's main interface uses standard windows icons and commands such as <u>N</u>ew, <u>O</u>pen, <u>S</u>ave, Cut, <u>C</u>opy, and <u>P</u>aste. Therefore, the environment outside the drawing area should look very familiar. It is the environment inside the drawing area or document window that may take a while to get used to. Shown below are the main components of CATIA's user interface followed by a description of each area.

1. <u>Pull down menus:</u> All of the commands, options, and settings may be accessed in the pull down menus.
2. <u>Active workbench icon:</u> CATIA has several workbench environments. For example, the *Sketcher* workbench is used to sketch 2D profiles and the *Part Design* workbench is used to make solids from a sketch among other things. The active workbench icon shows the workbench you are currently working in.

3. <u>Workbench specific toolbars:</u> Each workbench has its own specific set of toolbars. However, some toolbars are common between workbenches.

4. <u>Hidden toolbars:</u> If there is not enough room to show all active toolbars, CATIA will hide some. You will know when some toolbars are hidden if a double arrow appears in one or more of the corners.

5. <u>Prompt line:</u> The prompt line tells the user what the program is expecting next. For example, if the *Circle* command is chosen, the prompt line will read *Select a point or click to define the circle center.* After the center point is selected it will read *Select a point or click to define the circle radius.*

6. <u>Specification tree:</u> The specification tree shows all the commands and sketches that were used to construct the part. The specification tree may be used to make changes to the part.

7. <u>Compass:</u> The compass may be used to rotate and translate the environment or part.

8. <u>Coordinate planes:</u> The default coordinate planes show the 3 mutually perpendicular Cartesian coordinate planes. 2D profiles may be sketched on these planes.

9. <u>Active document window:</u> CATIA allows multiple documents or parts to be open simultaneously. These documents can be arranged in the usual windows based placements such as <u>T</u>iled horizontally and vertically, <u>C</u>ascaded, etc...

Mechanical Design Workbenches

Listed below are the most commonly used Mechanical Design workbenches along with a brief summary of their uses. The figure shows all of CATIA's Product lines and standard Mechanical Design workbenches.

- Part Design: The Part Design workbench allows you to create a main solid from a sketch and then modify it by adding or removing material.

- Sketcher: The Sketcher workbench is used to create 2D profiles that may be used to create the initial solid, modify an existing solid, or used as a path or direction curve.

- WireFrame & Surface: The WireFrame & Surface workbench is used to create complex wire frame profiles and surfaces.

- Assembly Design: The Assembly Design workbench is used to assemble and constrain parts in an assembly.

- Generative & Interactive Drafting: The Drafting workbench is used to create 2D detailed and assembly drawings from 3D models.

Documents

When working on a part or assembly in CATIA you are working within a document. The type of document depends on the workbench you are using to create and modify the geometry and each has its own file extension.

- Assembly (.CATProduct)
- Part Design, Sketcher, WireFrame & Surface (.CATPart)
- Interactive and Generative Drafting (.CATDrawing)

Shown on the next page, are 3 document windows open simultaneously in CATIA. There is an assembly of the product called *Machine Vise*, a part (the *Base*) belonging to the product and a 2D drawing of the *Base*.

Part Modeled

The part modeled in this tutorial is simple and is used to familiarize you with CATIA's interface.

Section 1: CATIA's Workbenches

1) Launch CATIA by double clicking on the **CATIA V5** icon on your desktop. (UNIX users: Check with you system administrator for the CATIA startup procedure.) Two windows will appear the outer *CATIA V5* window and the inner *Product* window.

CATIA V5R16

2) At the top pull down menu, select **Start – Mechanical Design – Part Design**. Workbenches may be accessed from the top pull down menu or from the workbench toolbar.

3) In the *New Part* window, name your part **Plate** and select **OK.** Maximize the *Part* window if it is not already maximized.

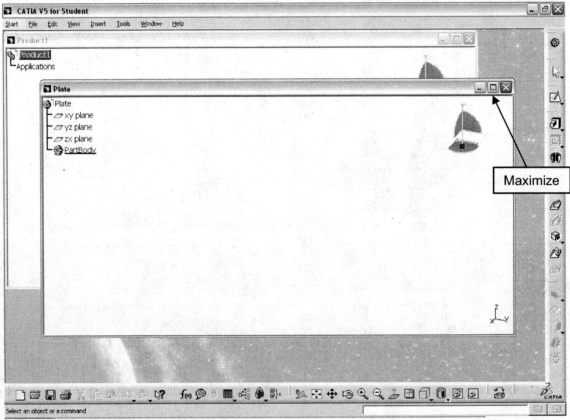

4) Create a list of favorite workbenches. At the top pull down menu, select **Tools** – **Customize.**

5) In the *Customize* window, select the **Start Menu** tab. Use the arrows to move the following workbenches from the *Available* area to the *Favorites* area: Assembly Design, Drafting, Part Design, Sketcher, and Wireframe and Surface Design. Then select the **Close** button.

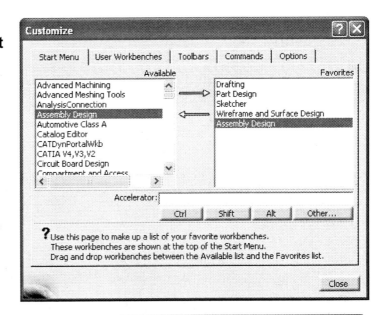

6) Select the **Part Design** workbench icon . A *Welcome to CATIA V5* window will appear that contains your favorite workbenches. This window will appear every time you start CATIA. **Close** the *Welcome to CATIA V5* window.

Section 2: The Specification Tree

2.1 – Specification Tree Basics

In the upper left corner of the *Part* window, there is a specification tree. This contains all the planes, sketches, operations, and constraints, used to create the part. The specification tree can be expanded and collapsed by clicking on the '+' and '-' symbols. The most important feature of the specification tree is that it allows you to view and modify the individual components and operations used to create the part. The specification tree is arranged in a hierarchical manner. At the top of the tree is the Part, under the Part there are 3 coordinate planes and an empty PartBody. The tree's structure will change depending on the type of document you are editing.

1) In the specification tree, click on the **xy plane**. Notice that the xy coordinate plane in the middle of the screen highlights in orange. Repeat for each coordinate plane.

When an item is selected in the specification tree, it is highlighted in orange.

CATIA's interface allows you to reposition toolbars to suit your needs. Therefore, they may not always be located in their original position. By restoring the toolbars default position, locating specific icons throughout each tutorial is easier.

2) Restore the toolbars to their default positions. At the top pull down menu, select **Tools – Customize...**

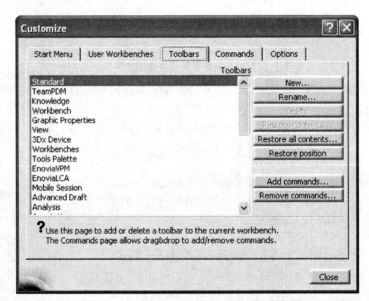

3) In the *Customize* window select the **Toolbars** tab and then select the **Restore position** button. In the *Restore all toolbars* window select **OK** and then select **Close** in the *Customize* window.

4) Enter the *Sketcher* Workbench on the xy plane. Select the **xy plane** in the specification tree or the coordinate plane itself. Then select the **Sketcher** icon

. This is usually located in the top right corner. In the *Sketcher* workbench a grid will appear and the environment will rotate such that you are viewing the xy plane.

5) Drag and drop the *User Selection Filter* and *Sketch Tools* toolbars to the top creating a top toolbar area. Place them in the approximate locations shown.

Locate the *Profile* toolbar in the right side toolbar area. Click on the bar at the top of this toolbar and drag it to the center of the screen. (If the *Profile* toolbar is not there, right click on the right side toolbar area and select Profile. A toolbar is active if there is a check mark next to it. UNIX users select from the top pull down menu, **View – Toolbars - Profile**)

Click on the bar to move toolbar.

Click on the black arrow to reveal hidden icons.

6) Click on the black arrow under the Rectangle icon to reveal the *Predefined Profile* toolbar.

7) Restore the toolbars default position as was done previously.

8) Drag and drop the *User Selection Filter* and *Sketch Tools* toolbars to the top.

9) Draw a Rectangle similar to the one shown. Use the grid as a guide. Select the **Rectangle** icon. It is usually located in the right side toolbar area. Notice that when you select the rectangle icon it turns orange. Down at the bottom of the screen the prompt line will read *Select or pick first point to create the rectangle*. Once you select the first point, the prompt line will read *Select or pick second point to create the rectangle*. The prompt line should be consulted whenever you are unsure what action to take next.

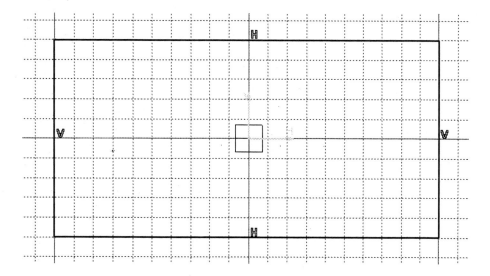

10) Exit the sketcher by selecting the **Exit workbench** icon .

11) Hit the **Esc** key twice or click on the graphics screen to deselect the rectangle. The rectangle should no longer be orange.

12) Pad the rectangle to a length of 20 mm. Select the **Pad** icon. In the *Pad Definition* window fill in the following fields; Type: **Dimension**, Length: **20 mm**. The Selection: field should read *No selection*. Activate the Selection: field by clicking on *No selection* and then click on the rectangle. The Selection: field should now read **Sketch.1**. Select **OK**. (UNIX users hold the CTRL key down when selecting the sketch.)

13) Create a 20 mm hole in the top of the block. Click on the **Hole** icon located in the right side toolbar area. Notice nothing happens. Read the prompt line. Select the top face of the block near the middle. In the *Hole Definition* window and **Extension** tab, fill in the following fields; select **Up To Last** in the top pull down field, Diameter: *20 mm*.

14) Expand your specification tree as shown. Notice that the PartBody in the specification tree is no longer empty. It contains a main solid (Pad.1) and a hole (Hole.1). Notice that each operation is based on a separate sketch.

Click on the vertical line to manipulate the tree.

2.2 – Manipulating the Specification Tree

The specification tree may be moved, zoomed or hidden.

1) Click on one of the vertical connecting lines in the tree. Your part should turn a dark grey this indicates that the tree is active. (The specification tree may also be activated and deactivated by using **Shift + F3**.)

2) Move the tree to the center of the screen by clicking on the vertical line again and dragging it.

3) Zoom in on your tree by clicking and holding your middle mouse button, clicking your right mouse button once, and moving your mouse up and down.

4) Move the tree back to its original location in the upper left corner and zoom it to its normal size.

5) Click on the vertical connecting line again to deactivate the tree.

6) Hit the **F3** key on your keyboard to hide the tree and then hit it again to show the tree.

Section 3: File Management

1) At the top pull down menu select **File – Save**. A Save As window will appear because this is the first save. Select the file location and name your file *Tutorial1-1* then select the **Save** button. The extension for a Part Design document is *CATPart*.

2) Enter the Sketcher on the top face of the block. Select the

   Sketcher icon and then select the top face of the block.

   Sketch face

   Face\Pad.1\PartBody

3) Draw a **Rectangle** in the approximate location and of approximate size as shown.

4) Exit the sketcher by selecting the **Exit workbench** icon .

5) **Pad** the rectangle to a length of *40 mm*.

6) At the top pull down menu select **File – Save**.

7) At the top pull down menu select **File – Close**. If a *Close* window appears, Select **No**.

8) At the top pull down menu select **File – New From...** Check the **Show Preview** box and click on **Tutorial 1-1.CATPart**. Select the **Open** button.

9) At the top pull down menu select **File – Save**. Notice that CATIA does not automatically save it under the original file name. Name the file *Tutorial 1-1b* and **Save** the file.

10) **Close** all documents.

11) At the top pull down menu select **File** – **New**.

12) In the *New* window, scroll down the options in the List of Types: window and select **Part** and then **OK**.

13) Select **OK** in the *New Part* window.

14) **Close** all documents.

Section 4: Using the Compass

The compass is used to translate and rotate the 3D environment and all of its components or an individual part.

1) Open *Tutorial1-1.CATPart*.

2) Click and hold one of the straight lines of the compass and move your mouse. Notice that both the part and the coordinate planes translate.

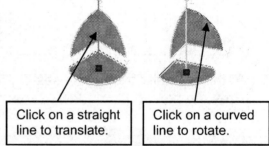

Click on a straight line to translate.

Click on a curved line to rotate.

3) Click and hold one of the curved lines of the compass and move your mouse. Notice that the part and the coordinate planes both rotate.

4) Click and hold on the top point of the compass and move your mouse. This allows you to rotate freely with no coordinate constraints.

Click on the top point to rotate freely.

5) Attach the compass to your part. Click and drag the red square of the compass and move it to the top of your part and release. The compass is attached to the part if it is green. If it is not green or it snaps back to it original position, move it again until it attaches to the part.

6) Repeat steps 1) and 2) and notice that the coordinate planes do not move. Only the part translates or rotates. If any warning windows popup informing you of constraint violations, click **OK** or **Yes**.

7) From the top pull down menu select **View** – **Reset Compass**.

8) Close without saving.

Chapter 1: CATIA BASICS

Tutorial 1.2:
Selecting, Editing & Viewing Objects

Featured Topics & Commands

Prerequisite Knowledge & Commands

- Entering workbenches

Simple Selection

The left mouse button is used to select an object. You can select an object or feature on the physical part or you can click on the corresponding feature in the specification tree. To select multiple items hold the control key down while selecting. The selected item(s) will be highlighted in orange in both the tree and the part.

The Selection Toolbar

There are several ways to select a feature. You can simply select the feature using your left mouse button or you can use one of the available traps. The *Selection* toolbar has 7 different selection options and is located near the top of the right side toolbar area. Reading the toolbar from left to right the options are;

- Select: This allows you to select features by using your left mouse button. To multiple select you need to hold the control key down. This option is always active.
- Selection Trap: The selection trap allows you to create a window and select all features that are completely enclosed in that window.
- Intersecting Trap: The intersecting trap uses a window to select all items that are enclosed in or touching the window boundaries.
- Polygon Trap: The polygon trap is similar to the selection trap but it enables you to create a window that is irregular.
- Paint Stroke Selection: The paint stroke selection allows you to create a freehand line. Everything that crosses this line is selected.
- Outside Trap Selection: The outside trap selection selects everything that is outside the window.
- Intersecting Outside Trap Selection: The intersecting outside trap selection selects everything that is outside and intersecting the window.

The View Toolbar

The *View* toolbar contains commands that allow you to manipulate (pan, rotate, zoom) your part and contains different viewing options. Reading the toolbar from left to right the options are;

- <u>Fly Mode:</u> Allows you to dynamically fly around the part. To use the fly mode you must be in a perspective view point (<u>V</u>iew – Rendering St<u>y</u>le – <u>P</u>erspective)
- <u>Fit All In:</u> This zooms your part out or in to fit the available window space.
- <u>Pan:</u> Allows you to move your part around the screen without zooming.
- <u>Rotate:</u> Allows you to rotate your part to view different sides.
- <u>Zoom In:</u> Makes your part look bigger.
- <u>Zoom Out:</u> Makes your part look smaller.
- <u>Normal View:</u> It rotates the part to show you a 2D view of the object relative to a selected face.
- <u>Create Multi-view:</u> This command allows you to create four different independently controlled views of the part and/or sketch.
- <u>Quick view toolbar:</u> Allows you to view the part from the following different perspectives.
  - Isometric view
  - Front view
  - Back view
  - Left view
  - Right view
  - Top view
  - Bottom view
  - Named views

- <u>View mode toolbar:</u> Enables you to view the part in the following different modes.
  - Shading (SHD)
  - Shading with Edges
  - Shading with Edges without Smooth Edges
  - Shading with Edges and Hidden Edges
  - Shading with Material
  - Wireframe (NHR)
  - Customize View Parameters

- <u>Hide/Show:</u> Allows you to hide and show different features of a part.
- <u>Swap visible space:</u> This command hides everything that is showing and shows everything that is hidden.

The Graphics Properties Toolbar

The *Graphic Properties* toolbar allows you to change the fill color, line thickness and type, and transparency of an object among other things. The options in the *Graphic Properties* toolbar from left the right are;

- <u>Fill Color:</u> Allows you to change the color of a part.

- <u>Transparency:</u> You can make a part transparent so that you can see parts or features behind it.
- <u>Line Thickness:</u> You can change the line thickness of a sketch.
- <u>Line Type:</u> You can change the line type of a sketch.
- <u>Point Style:</u> There are several different choices for point style.
- <u>Painter:</u> This allows you to match the color of one feature to the color of another.

Part Modeled

The part modeled in this tutorial is shown on the right. It will be used to illustrate how to edit, manipulate, view and change the graphics properties of a part.

Section 1: Selecting profiles and features

1) Launch **CATIA V5**, enter the **Part Design** workbench, and name your part *Seat*.

2) Enter the Sketcher workbench on the xy plane. Select the **xy plane** in the specification tree or the coordinate plane it self then select the **Sketcher** icon.

3) Turn the snap to grid option on. Select the **Snap to Point** icon. If the icon is orange, the snap is on, if it is blue, the snap is off. The *Snap to Point* icon is located in the *Sketch Tools* toolbar. This toolbar should be located at the top toolbar area where it was moved. If this toolbar is not active or you can't locate it, go to **Tools – Customize... – Toolbar** tab – **Restore Position**.

4) Select the **Profile** icon and draw the sketch shown. The profile command allows you to draw several continuous lines connected together. Count the grid spacing to determine the sketch size.

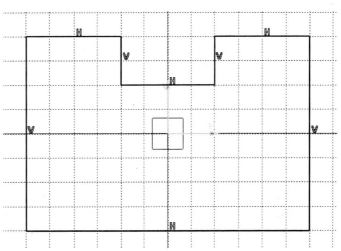

5) Click on the screen somewhere away from the sketch to deselect the profile. Nothing should be highlighted in orange.

6) Click on the left vertical line of the profile. Notice it turns orange.

7) Click on the bottom line in the profile. Notice that the bottom line is orange, but the vertical line is no longer orange.

8) Hold the control key and click on the right side vertical line of the profile. Notice that the bottom line remains selected.

9) Deselect all. Click off of the profile or hit the Esc key twice.

10) Bring the **Select toolbar** out from underneath the **Select** icon and move it so that it stands alone.

11) In the *Select* toolbar, select the **Selection Trap** icon. Draw the selection window shown by clicking and holding the mouse button and moving to the opposite corner. Notice that the only the lines that where entirely enclosed in the window are selected.

Selection window

12) Deselect all.

13) In the *Select* toolbar, select the **Intersecting Trap** icon. Draw a selection window like the previous window. Notice that this time all items that where enclosed in and touching the window are selected.

14) Deselect all.

15) In the *Select* toolbar, select the **Polygon Trap** icon. Draw the selection window shown clicking the mouse at positions 1 to 4 in order. Double click at point 4 to end the window selection.

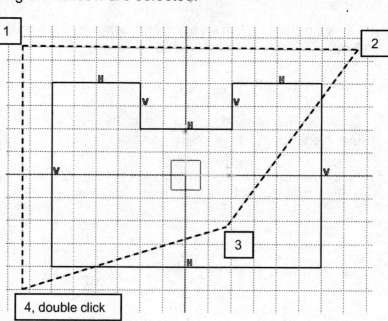

1

2

3

4, double click

16) Deselect all.

17) Try the **Outside Trap Selection** icons and determine what they do.

18) Use the **Selection Trap** to select the entire sketch.

19) Hold the **control** key and deselect the bottom line by clicking on it.

20) Close the **Select toolbar**.

21) Exit the sketcher by selecting the **Exit workbench** icon.

22) Pad the profile to a length of 20 mm. Select the **Pad** icon. In the *Pad Definition* window fill in the following fields; Type: **Dimension**, Length: *20 mm*, Selection: **Sketch.1**.

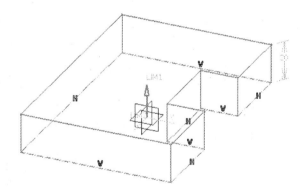

Section 2: Undo/Redo

1) Enter the Sketcher workbench on the top face of the part. Select the **Sketcher** icon and then select the top face of the part.

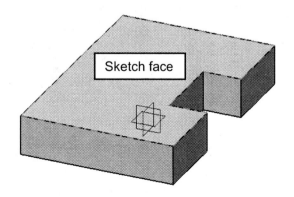

Sketch face

2) Draw the following **Profile** .

3) **Exit** the Sketcher and **Pad** the sketch to a length of **50 mm**.

4) Enter the **Sketcher** on the **zx plane**.

5) Draw the Line shown in white.

6) **Exit** the Sketcher.

7) Select the **Stiffener** icon. This is located in the *Sketch Based Features* toolbar. It may be stacked under the **Solid Combine** icon. Create a stiffener using **Sketch.3** that is *1 mm* thick and is **From the Side**.

8) Enter the **Sketcher** on the front face of the part.

Sketch face

9) Draw 2 **Rectangles** as shown.

10)**Exit** the Sketcher.

11) Create pockets using the two rectangles. The Pocket command removes material. Select the

Pocket icon and use the **Up to Last** option (this means through all).

12) Expand your specification tree to the sketch level as shown.

13) Select **Stiffener.1** in the specification tree. Notice that this feature is highlighted in the specification tree as well as in the part.

14) Right click on **Stiffener.1** and select **Delete**. Select **OK** in the *Delete* window.

15) At the top pull down menu, select **Edit – Undo**.

Alternatively, you can select the **Undo** icon, or use the keyboard shortcut **CTRL + Z**.

16) Right click on **Sketch.3** (the sketch associated with the stiffener) and select **Hide/Show**. Now we can see the sketch used to create the stiffener.

17) Select the **Undo with History** icon. This is stacked under the *Undo* icon. Your undo history should look similar to the one shown. There may be some extra steps depending on what you have done. Select **(Pocket)**. This will select all commands performed before the Pocket as well as the Pocket command. Select the **Undo** icon and then **Close**.

18) At the top pull down menu, select **Edit** – **Repeat**, select the **Redo** icon, or use the keyboard shortcut **CTRL + Y** to get the Pocket back.

Section 3: Editing features using the compass

1) Deselect all.

2) Click and drag the red square of the compass to the top surface of your original pad as shown. Notice that Pad.1 highlights in the specification tree.

3) In the specification tree, click on **Stiffener.1**. Now the compass is attached to the stiffener instead of the pad.

4) Right click on the compass and select **Edit...**

5) In the *Parameters for Compass Manipulation* window, enter **10 mm** in each of the *Translation increment* fields (U, V & W) and select **Close**. By doing this you are setting the minimum increment that the compass is allowed to move.

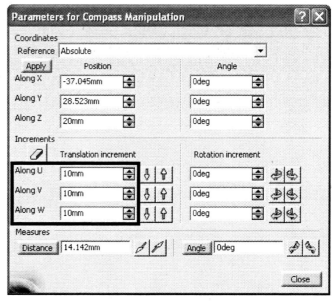

6) Move the compass one increment in the Y or V direction and one increment in the Z or W direction. When the *Move Warning* window appears select **Yes** or **OK**. (Hint: To translate the compass, click and drag on one of the straight axis lines.) When complete your part should look like the figure shown.

7) At the top pull down menu, select **View – Reset Compass**.

Section 4: Scanning the design process

This allows you to view the part's history step by step in order to see the design process.

1) At the top pull down menu, select **Edit – Scan or Define In Work Object...**

2) In the *Scan* window, select the **First** button, then scroll through the design process using the **Next** button. When complete, select the **Exit** button.

Section 5: Viewing your part

The part can be manipulated (Pan, Rotate, Zoom) using the icons in the *View* toolbar or by using the mouse buttons. Given below is a description of how to use a three button mouse to manipulate your part.

- Pan = Hold the middle mouse button and move the mouse.
- Zoom = Hold the middle mouse button, click and release the right mouse button and move the mouse.
- Rotate = Hold the middle mouse button, click and hold the right or left mouse button and move the mouse.

1) Use the icons in the *View* toolbar to **Pan** ⊕, **Rotate** ↩, and **Zoom** ⊕ ⊖ your part. This toolbar is located in the bottom toolbar area. It may be hidden in the bottom right corner.

2) Use your mouse buttons to Pan, Rotate and Zoom your part.

3) Select the **Fit All In** icon.

4) View your object from the **Top view**, then the **Right view**, and then the **Isometric view**.

5) Select different view modes and notice the differences. Go back to **Shaded with Edges** when finished.

Section 6: Graphic Properties

Parts in CATIA can be assigned different colors. This is very helpful especially when viewing an assembly.

1) In your specification tree, right click on **PartBody** and select **Properties**.

2) In the *Properties* window, select the *Graphic* tab. Under the *Fill Color* field select the color **red** and select **Apply**. Notice that the entire part changes to red. Then, under the *Edges Color* field select the color **green** and select **OK**.

3) Deselect all.

4) In your specification tree, right click on **Stiffener.1** and select **Properties**. Set the fill color to **tan**.

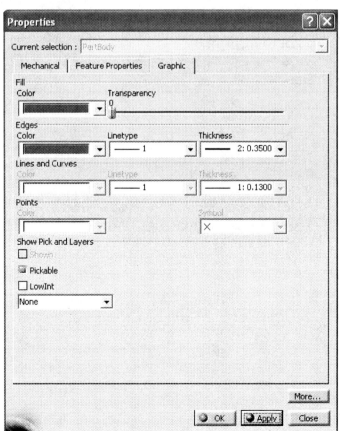

5) At the top pull down menu, select **View – Toolbars – Graphic Properties**. The *Graphics Properties* toolbar should appear in the top toolbar area.

6) In the specification tree, select **Pad.2** and change its fill color to **light blue** using the *Graphic Properties* toolbar.

Fill color   Transparency

7) In the specification tree, select **Stiffener.1** and change its Transparency to **50%** using the *Graphic Properties* toolbar.

8) Using the **control** key, pick both surfaces that were used to sketch the pocket shapes as shown in the figure and change there color to **light blue**.

9) Deselect all.

10) Hold **control key** down and select **Pad.1** in the specification tree. Still holding the control key, select the

Painter icon and select **Pad.2**. This procedure matches the color of Pad.1 to Pad.2.

Select these surfaces.

11) Using the **Painter** icon, change the color of the **Pocket.1** to make the color of **Pad.1**.

Chapter 2: SKETCHER

Introduction

Chapter 2 focuses on CATIA's *Sketcher* workbench. The reader will learn how to sketch and constrain very simple to very complex 2D profiles.

Tutorials Contained in Chapter 2

- Tutorial 2.1: Sketch Work Modes
- Tutorial 2.2: Simple Profiles & Constraints
- Tutorial 2.3: Advanced Profiles & Sketch Analysis
- Tutorial 2.4: Modifying Geometries & Relimitations
- Tutorial 2.5: Axes & Transformations
- Tutorial 2.6: Operations on 3D Geometries & Sketch planes
- Tutorial 2.7: Points & Splines

NOTES:

Chapter 2: SKETCHER

Tutorial 2.1: Sketch Work Modes

Featured Topics & Commands

Prerequisite Knowledge & Commands

- Entering workbenches
- Entering and exiting the *Sketcher* workbench
- Drawing simple profiles
- Simple Pads and Pockets

The Sketcher Workbench

The *Sketcher* workbench is a set of tools that helps you create and constrain 2D geometries. Features (pads, pockets, shafts, etc...) may then be created solids or modifications to solids using these 2D profiles. You can access the *Sketcher* workbench in various ways. Two simple ways are by using the top pull down menu (<u>S</u>tart – <u>M</u>echanical Design – <u>S</u>ketcher), or by selecting the *Sketcher* icon. When you enter the sketcher, CATIA requires that you choose a plane to sketch on. You can choose this plane either before or after you select the *Sketcher* icon. To exit the sketcher, select the *Exit Workbench* icon.

The *Sketcher* workbench contains the following standard workbench specific toolbars.

- <u>Profile toolbar:</u> The commands located in this toolbar allow you to create simple geometries (rectangle, circle, line, etc...) and more complex geometries (profile, spline, etc...).

- <u>Operation toolbar:</u> Once a profile has been created, it can be modified using commands such as trim, mirror, chamfer, and other commands located in the *Operation* toolbar.

- <u>Constraint toolbar:</u> Profiles may be constrained with dimensional (distances, angles, etc...) or geometrical (tangent, parallel, etc...) constraints using the commands located in the *Constraint* toolbar.

- <u>Sketch tools toolbar:</u> The commands in this toolbar allow you to work in different modes which make sketching easier.

- <u>User Selection Filter toolbar:</u> Allows you to activate different selection filters.

- Visualization toolbar: Allows you to, among other things to cut the part by the sketch plane and choose lighting effects and other factors that influence how the part is visualized.

- Tools toolbar: Allows you to, among others other things, to analyze a sketch for problems, and create a datum.

The Sketch tools Toolbar

The *Sketch tools* toolbar contains icons that activate and deactivate different work modes. These work modes assist you in drawing 2D profiles. Reading from left to right, the toolbar contains the following work modes; (Each work mode is active if the icon is orange and inactive if it is blue.)

- Grid: This command turns the sketcher grid on and off.
- Snap to Point: If active, your cursor will snap to the intersections of the grid lines.

- Construction / Standard Elements: You can draw two different types of elements in CATIA a standard element and a construction element. A standard element (solid line type) will be created when the icon is inactive (blue). It will be used to create a feature in the *Part Design* workbench. A construction element (dashed line type) will be created when the icon is active (orange). They are used to help construct your sketch, but will not be used to create features.
- Geometric Constraints: When active, geometric constraints will automatically be applied such as tangencies, coincidences, parallelisms, etc...
- Dimensional Constraints: When active, dimensional constraints will automatically be applied when corners (fillets) or chamfers are created, or when quantities are entered in the value field. The value field is a place where dimensions such as line length and angle are manually entered.

Part Modeled

The part modeled in this tutorial is shown below. The part is constructed with the assistance of different work modes.

Section 1: Using Snap to Point

1) Open a **New Part** drawing and name the part **Spline Shape**.

2) Enter the **Sketcher** on the **yz plane**.

3) Restore the default positions of the toolbars (**Tools – Customize... – Toolbars** tab – **Restore position**.) Move the *Sketch Tools* toolbar and the *User Selection Filter* toolbar to the top toolbar area.

4) Set your grid spacing. At the top pull down menu, select **Tools – Options...** In the *Options* window, expand the Mechanical Design portions of the left side navigation tree and select **Sketcher**. Activate the options *Display, Snap to point,* and *Allow Distortions* in the Grid section on the right side. Set your Primary spacing and Graduations to H: ***100 mm*** and ***20***, and V: ***100 mm*** and ***10***.

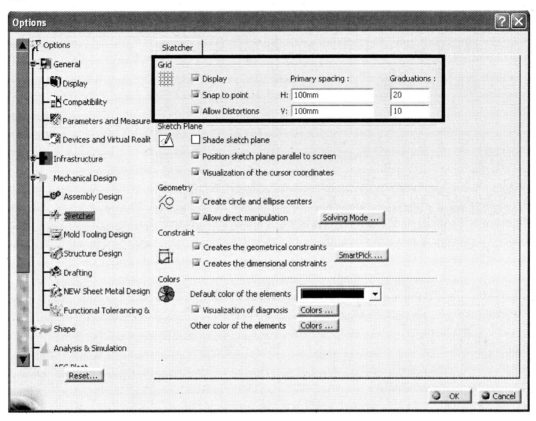

5) Select the **Spline** icon. This is located in the *Profile* toolbar in the right side toolbar area.

6) Move your cursor around the screen. Note that it snaps to the intersections of the grid. Your *Snap to Point* should be orange (active). Deactivate the **Snap to Point** icon by clicking on it and turning it back to blue. Move your cursor around the screen and notice the difference.

7) Reactivate the **Snap to Point** 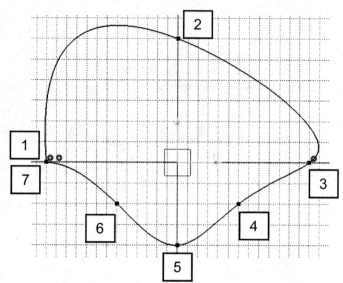 icon and draw the spline shown. Select each point (indicated by a number in a square) in order from 1 to 7, double clicking at the last point to end the spline command.

8) Edit the spline by double clicking on any portion of it.

9) In the *Spline Definition* window, select **CtrlPoint.7**, then activate the **Tangency** option, and select **OK**. Notice that the last point is now tangent to the first point.

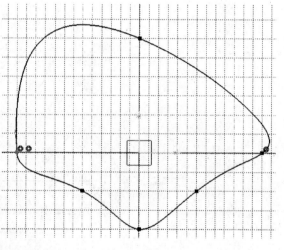

10) Draw a **Circle** inside the spline as shown.

11) **Exit** the Sketcher and **Pad** the sketch to a length of **50 mm**.

Section 2: Using construction elements.

1) Deselect all.

2) Enter the **Sketcher** on the front face of the part.

3) Activate the **Construction / Standard Elements** icon. It should be orange.

4) Deselect all. Hit the **Esc** key twice.

5) Project an outline of the part onto the sketch plane. Select the **Project 3D Elements** icon then select the face of the part. This icon is located in the *Operations* toolbar near the bottom of the right side toolbar area. It may be hidden in the bottom right corner.

6) Deselect all. The projection should now be yellow (this means it is associated with the part and will change with the part) and dashed (this means it is a construction element).

7) At the top pull down window, select **Tools** – **Options** – **Sketcher** tab. Deactivate the Grid **Display** and **Snap to Point** options. Select **OK**.

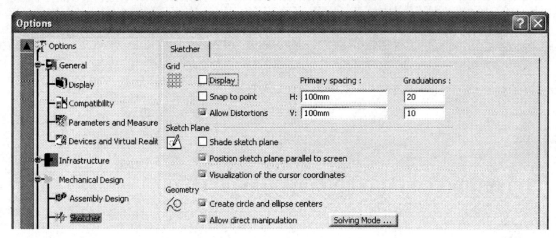

8) Deactivate the **Construction / Standard Elements** icon.

9) Using the **Profile** command to draw the triangle shown. The points of the triangle should lie on the projected construction element. You will know when you are on the projection when a symbol of two concentric circles appears, and you will know when you are snapped to the endpoint of the start point when a symbol of two concentric circles appears and the inner one is filled.

10) **Exit** the Sketcher and **Pad** the sketch to a length of **10 mm**.

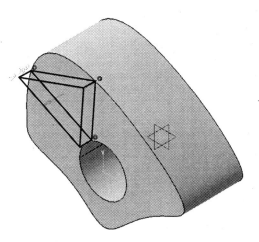

Section 3: Geometrical and Dimensional Constraints

1) Deselect all.

2) Enter the **Sketcher** on the front large face of the part.

3) Activate the **Geometrical Constraints** icon. It should be orange.

Sketch face

4) At the top pull down window, select **Tools** – **Options** – **Sketcher**. Under the Constraint portions of the window, select **SmartPick...** The *SmartPick* window shows all the geometrical constraints that will be created automatically. These constraints may be turn on and off depending on your design/sketch needs. **Close** both the *Smart Pick* and *Options* windows.

5) Draw a **Rectangle** to the right of the hole as shown. Notice that geometric constraints (H = horizontal, V = Vertical) are automatically applied.

6) Deactivate the **Geometrical Constraints** icon. It should be blue.

7) Draw a **Rectangle** to the left of the hole as shown. Notice that no geometric constraints are made.

Click and drag the corner point.

8) For each rectangle, click on one of the points defining a corner and move it using the mouse. Notice the difference between the two. This is due to the horizontal and vertical constraints that were applied to the one rectangle.

9) Undo (**CTRL + Z**) the moves until the original rectangles are back.

10) **Exit** the Sketcher and **Pocket** the sketch using the **Up to last** option.

11) Expand the specification tree to the sketch level.

12) Edit Sketch.3 (the sketch associated with the pocket). In the specification tree, double click on **Sketch.3**, or right click on it and select **S̲ketch.3 object - E̲dit**. You will automatically enter the sketcher on the sketch plane used to create this sketch.

13) Activate the **Dimensional Constraint** icon. It should be orange.

14) Select the **Corner** icon, select the bottom left corner point of the left rectangle, move your mouse up and to the right, and click. A corner or fillet will be created. The corner icon is located in the *Operations* toolbar near the bottom of the right side toolbar area. The corner/fillet may also be created by selecting the two lines that create the corner. Notice that a dimension is automatically created.

Corner point

15) Deactivate the **Dimensional**

Constraint icon. It should be

blue. Create a **Corner** in the upper right corner of the same rectangle. Notice that this time no dimensional constraint was created.

16) **Exit** the Sketcher . We have changed the sketch used to create the pocket. Notice that the pocket is automatically updated to reflect these changes.

Section 4: Cutting the part by the sketch plane.

Sometimes it is necessary to sketch inside the part. The *Cut Part by Sketch Plane* command allows you to see inside the part and makes it easier to draw and constrain your sketch.

1) Enter the **Sketcher** on the **xy plane**.

2) Select the **Isometric View** icon. This icon is located in the bottom toolbar area.

3) Select the **Cut Part by Sketch Plane** icon located in the bottom toolbar area. The part in now cut by the xy plane (the sketch plane).

4) Select the **Top view** icon and draw a **Circle** in the middle of the hole as shown in the figure.

5) **Exit** the Sketcher.

6) Select the **Pad** icon and then select the **More>>** button. Fill in the following fields for both the First and Second Limits; Type: **Up to surface**, Limit: Select the inner circumference of the hole, and Selection: **Sketch.4** (the circle). Select **Preview** to see if the Pad will be applied correctly, and then **OK**.

Pad Definition

First Limit
Type: Up to surface
Limit: Pocket.1\Face.1
Offset: 0mm

Second Limit
Type: Up to surface
Limit: Pocket.1\Face.2
Offset: 0mm

Profile/Surface
Selection: Sketch.4
☐ Thick
Reverse Side

☐ Mirrored extent
Reverse Direction

Direction
☑ Normal to profile
Reference: No selection

Thin Pad
Thickness1: 1mm
Thickness2: 0mm
☐ Neutral Fiber ☐ Merge Ends

<<Less

OK   Cancel   Preview

Chapter 2: SKETCHER

Tutorial 2.2: Simple Profiles & Constraints

<u>Featured Topics & Commands</u>

<u>Prerequisite Knowledge & Commands</u>

- Entering workbenches
- Entering and exiting the *Sketcher* workbench
- Simple Pads
- Work modes (Sketch tools toolbar)

Profile toolbar

The *Profile* toolbar contains 2D geometry commands. These geometries range from the very simple (point, rectangle, etc...) to the very complex (splines, conics, etc...). The *Profile* toolbar contains many sub-toolbars. Most of these sub-toolbars contain different options for creating the same geometry. For example, you can create a simple line, a line defined by two tangent points, or a line that is perpendicular to a surface. Reading from left to right, the *Profile* toolbar contain the following commands.

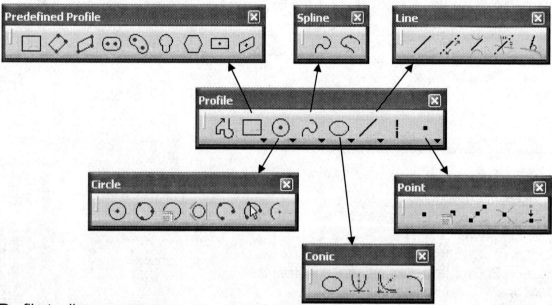

Profile toolbar

- Profile: This command allows you to create a continuous set of lines and arcs connected together.
- Rectangle / *Predefined Profile* toolbar: The default top command is *rectangle*. Stacked underneath are several different commands used to create predefined geometries.
- Circle / *Circle* toolbar: The default top command is *circle*. Stacked underneath are several different options for creating circles and arcs.
- Spline / *Spline* toolbar: The default top command is *spline* which is a curved line created by connecting a series of points.
- Ellipse / *Conic* toolbar: The default top command is *ellipse*. Stacked underneath are commands to create different conic shapes such as a hyperbola.
- Line / *Line* toolbar: The default top command is *line*. Stacked underneath are several different options for creating lines.
- Axis: An axis is used in conjunction with commands like mirror and shaft (revolve). It defines symmetry. It is a construction element so it does not become a physical part of your feature.

- Point / *Point* toolbar: The default top command is *point*. Stacked underneath are several different options for creating points.

Predefined Profile toolbar

Predefined profiles are frequently used geometries. CATIA makes these profiles available for easy creation which speeds up drawing time. Reading from left to right, the *Predefined Profile* toolbar contains the following commands.

- Rectangle: The *rectangle* is defined by two corner points. The sides of the rectangle are always horizontal and vertical.
- Oriented Rectangle: The *oriented rectangle* is defined by three corner points. This allows you to create a rectangle whose sides are at an angle to the horizontal.
- Parallelogram: The *parallelogram* is defined by three corner points.
- Elongated Hole: The elongated hole or slot is defined by two points and a radius.
- Cylindrical Elongated Hole: The *cylindrical elongated hole* is defined by a cylindrical radius, two point and a hole radius.
- Keyhole Profile: The *keyhole profile* is defined by two center points and two radii.
- Hexagon: The *hexagon* is defined by a center point and the radius of an inscribed circle.
- Centered Rectangle: The *centered rectangle* is defined by a center point and a corner point.
- Centered Parallelogram: The *centered parallelogram* is defined by a center point (defined by two intersecting lines) and a corner point.

Circle toolbar

The *Circle* toolbar contains several different ways of creating circles and arcs. Reading from left to right, the *Circle* toolbar contains the following commands.

- Circle: A *circle* is defined by a center point and a radius.
- Three Point Circle: The *three point circle* command allows you to create a circle using three circumferential points.
- Circle Using Coordinates: The *circle using coordinates* command allows you to create a circle by entering the coordinates for the center point and radius in a *Circle Definition* window.
- Tri-Tangent Circle: The *tri-tangent circle* command allows you to create a circle whose circumference is tangent to three chosen lines.

- Three Point Arc: The *three point arc* command allows you to create an arc defined by three circumferential points.
- Three Point Arc Starting With Limits: The *three point arc starting with limits* allows you to create an arc using a start, end, and midpoint.
- Arc: The *arc* command allows you to create an arc defined by a center point, and a circumferential start and end point.

Spline toolbar

Reading from left to right, the *Spline* toolbar contains the following commands.

- Spline: A spline is a curved profile defined by three or more points. The tangency and curvature radius at each point may be specified.
- Connect: The connect command connects two points or profiles with a spline.

Conic toolbar

Reading from left to right, the *Conic* toolbar contains the following commands.

- Ellipse: The ellipse is defined by center point and a major and minor axis points.
- Parabola by Focus: The parabola is defined by a focus, apex and a start and end point.
- Hyperbola by Focus: The hyperbola is defined by a focus, center point, apex and a start and end point.
- Conic: There are several different methods that can be used to create conic curves. These methods give you a lot of flexibility when creating above three types of curves.

Line toolbar

The *Line* toolbar contains several different ways of creating lines. Reading from left to right, the *Line* toolbar contains the following commands.

- Line: A line is defined by two points.
- Infinite Line: Creates infinite lines that are horizontal, vertical or defined by two points.
- Bi-Tangent Line: Creates a line whose endpoints are tangent to two other elements.
- Bisecting Line: Creates an infinite line that bisects the angle created by two other lines.
- Line Normal to Curve: This command allows you to create a line that starts anywhere and ends normal or perpendicular to another element.

Point toolbar

The *Point* toolbar contains several different ways of creating points. Reading from left to right, the *Point* toolbar contains the following commands.

- <u>Point by Clicking:</u> Creates a point by clicking the left mouse button.
- <u>Point by using Coordinates:</u> Creates a point at a specified coordinate point.
- <u>Equidistant Points:</u> Creates equidistant points along a predefined path curve.
- <u>Intersection Point:</u> Creates a point at the intersection of two different elements.
- <u>Projection Point:</u> Projects a point of one element onto another.

Constraint toolbar

Constraints can either be dimensional or geometrical. Dimensional constraints are used to constrain the length of an element, the radius or diameter of an arc or circle, and the distance or angle between elements. Geometrical constraints are used to constrain the orientation of one element relative to another. For example, two elements may be constrained to be perpendicular to each other. Other common geometrical constraints include parallel, tangent, coincident, concentric, etc... Reading from left to right:

- <u>Constraints Defined in Dialoged Box:</u> Creates geometrical and dimensional constraints between two elements.
- <u>Constraint:</u> Creates dimensional constraints.
- <u>Contact Constraint:</u> Creates a contact constraint between two elements.
- <u>Fix Together:</u> The fix together command groups individual entities together.
- <u>Auto Constraint:</u> Automatically creates dimensional constraints.
- <u>Animate Constraint:</u> Animates a dimensional constraint between to limits.
- <u>Edit Multi-Constraint:</u> This command allows you to edit all your sketch constraints in a single window.

Selecting icons

When an icon is selected, it turns orange indicating that it is active. If the icon is activated with a single mouse click, the icon will turn back to blue (deactivated) when the operation is complete. If the icon is activated with a double mouse click, it will remain active until another command is chosen or if the Esc key is hit twice.

Part Modeled

The part modeled in this tutorial is shown on the right. This part will be created using simple profiles, circles, arcs, lines, and hexagons. The geometries are constrained to conform to certain dimensional (lengths) and geometrical constraints (tangent, perpendicular, etc...).

Section 1: Creating circles.

(Hint: If you get confused about how to apply the different commands that are used in this tutorial, read the prompt line for additional help.)

1) Launch **CATIA V5,** enter the **Part Design** workbench and, if asked, name your part **Post**.

2) Enter the **Sketcher** on the **zx plane**.

3) Set your grid spacing to be **100 mm** with **10** graduations, activate the *Snap to point*, and activate the *geometrical* and *dimensional* constraints. (**Tools – Options...**)

4) Pull out the **Circle** toolbar .

5) Double click on the **Circle** icon and draw the circles shown.

6) **Exit** the Sketcher and **Pad** the sketch to **12 mm** on each side (**Mirrored extent**). Notice that the inner circle at the bottom becomes a hole.

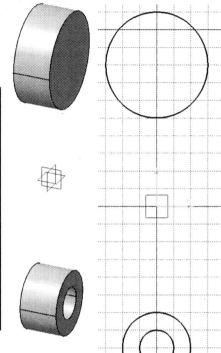

Section 2: Creating dimensional constraints.

1) Expand your specification tree to the sketch level.

2) Edit Sketch.1. To edit a sketch you can double click on the sketch name in the specification tree, or you can *right click* on the name select **Sketch.1 - Edit**. CATIA automatically takes you into the sketcher on the plane used to create Sketch.1.

3) Double click on the **Constraints** icon.

4) Select the border of the upper circle, pull the dimension out and click your left mouse button to place the dimension. Repeat for the two bottom circles.

5) Select the center point of the upper circle, then the center point of the lower circles, pull the dimension out and click.

6) Double click on the **D20** dimension. In the *Constraint Definition* window, change the diameter from 20 to **16 mm**.

7) In a similar fashion, change the other dimensions to the values shown in the figure.

8) **Exit** the Sketcher 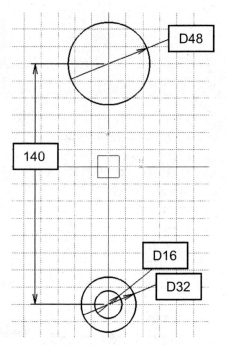 and deselect all. Notice that the part automatically updates to the new sketch dimensions.

Section 3: Creating lines.

1) Enter the **Sketcher** on the **zx plane**.

2) Deactivate the **Snap to Point** icon.

3) Project the two outer circles of the part onto the sketch plane. Double click on the **Project 3D Elements** icon. This icon is located in the lower half of the right side toolbar area. Select the outer edges of the two cylinders.

4) Pull out the *line* toolbar.

5) Double click on the **Bi-Tangent Line** icon. Select the points, in order, as indicated on the figure.

6) Pull out the **Relimitations** toolbar located in the *Operation* toolbar.

7) Double click on the **Quick trim** icon. Select the outer portion of the projected circles. Notice that the trimmed projection turns into a construction element (dashed).

8) **Exit** the Sketcher and **Pad** the sketch to *6 mm* on each side (**Mirrored extent**).

Pad Definition

First Limit

Type: Dimension

Length: 6mm

Limit: No selection

Profile/Surface

Selection: Sketch.2

☐ Thick

Reverse Side

☑ Mirrored extent

Reverse Direction

More>>

OK   Cancel   Preview

9) Enter the **Sketcher** 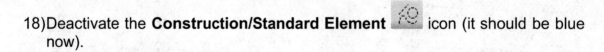 on the **zx plane**.

10) Activate the **Construction/Standard Element** icon (it should be orange).

11) Select the **Project 3D Elements** icon and then project the left line of the part as shown in the figure.

12) Activate your **Snap to Point** icon.

13) Draw a line that starts at point 1 (see fig.) and ends normal/perpendicular to projected line using the **Line Normal to Curve** icon.

14) Deactivate your **Snap to Point** icon.

15) Draw a **Line** from point 1 to point 2.

16) Draw a line that bisects the previous 2 lines using the **Bisecting Line** icon. Read the prompt line for directions.

17) Deselect all.

18) Deactivate the **Construction/Standard Element** icon (it should be blue now).

Projected line

Normal line

1

Bisecting line

2

19) Draw a circle that is tangent to the projected line, normal line and bisecting line using the **Tri-Tangent Circle** icon. Read the prompt line for directions.

20) Zoom in on the circle.

21) Using **Profile**, draw the three additional lines shown in the figure.

22) Use the **Quick Trim** command to trim off the inside portion of the circle as shown. You will have to apply the quick trim operation twice.

23) Draw a **Hexagon** that has the same center as the circle/arc and is the approximate size shown in the figure. The *Hexagon* icon is usually stacked under the *Rectangle* icon. (Your hexagon will contain many constraints that are not shown in the figure.)

24) Deselect all.

25) Apply a dimensional **Constraint** to the distance between the flats of the hexagon as shown. To create this constraint, select the top line and then the bottom line. Double click on the dimension and change its value to **7 mm**.

26) **Exit** the Sketcher and **Pad** the sketch to a length of **2 mm** on each side (**Mirrored extent**).

Section 4: Creating geometrical constraints.

1) Enter the **Sketcher** 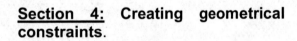 on the flat face of the large cylinder.

Sketch face

2) Deactivate the **Geometrical Constraint** 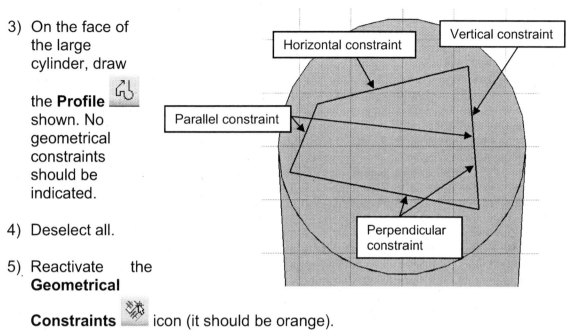 icon (it should be blue). This will allow you to create profiles with no automatically applied constraints.

3) On the face of the large cylinder, draw the **Profile** shown. No geometrical constraints should be indicated.

4) Deselect all.

5) Reactivate the **Geometrical Constraints** icon (it should be orange).

6) Apply a vertical constraint to the right line of the profile by right clicking on it and selecting **L̲ine.? object – V̲ertical**.

7) Apply a horizontal constraint to the top line using a similar procedure.

8) Deselect all.

9) Apply a perpendicular constraint between the right and bottom line of the profile. Hold the CTRL key down and select the left and bottom lines. Select the **Constraints Defined in Dialog Box** icon. In the *Constraint Definition* window, check the box next to **Perpendicular** and then select **OK.**

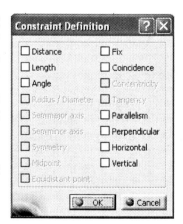

10) Apply a parallel constraint between the left and right lines of the profile in a similar way.

11) Apply **Constraints** to the rectangle and change their values to the values shown in the figure.

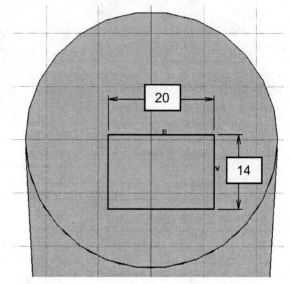

12) Apply the additional dimensional constraints shown in order to position the rectangle. Select the

Constraints icon, then the circumference of the circle and then the appropriate side of the rectangle. Notice that once all the constraints are applied, the rectangle turns green indicating that it is fully constrained. If it did not turn green make sure the *Visualization of diagnosis* is activated in the *Options* window. (Tools – Options...)

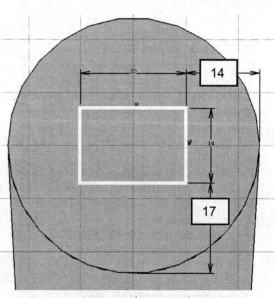

13) Draw the triangle shown using the

Profile command. When drawing the triangle make sure that the top point is aligned with the origin ( ) and the bottom line is horizontal (H).

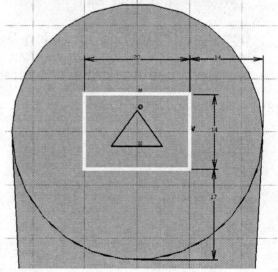

14) Constrain the vertical height of the triangle to be **6 mm**. Select the **Constraints** icon, select the one of the angled lines of the triangle, right click and select **V**ertical Measure Direction.

15) **Constrain** the rest of the triangle as shown.

16) **Exit** the Sketcher and **Pad** the sketch to a length of **5 mm.**

Section 5: Creating arcs.

1) Enter the **Sketcher** on the front face of the middle section.

Sketch face

2) Activate the **Construction/Standard Element** icon.

3) Select the **Project 3D Elements** icon and then project the front face of the middle section.

4) Deselect all.

5) Deactivate the **Construction/Standard Element** icon.

6) Activate your **Snap to Point** icon.

7) Draw the profile shown. Use the **Three Point Arc** command to create the bottom arc, the **Arc** command to create the top arc. The *Arc* icons are stacked under the *Circle* icon. For assistance in creating the arcs, read the prompt line at the bottom of the graphics screen. Use the **Profile** command to create the connecting lines.

Center point for arc

Arc

Three point arc

8) **Exit** the Sketcher and **Pad** the sketch to a length of **30 mm.**

9) Deselect all.

10) Mirror the entire solid. Select the **Mirror** icon in the Transformation Features toolbar. Select the mirror element/face. In the Mirror Definition window select **OK**.

Mirroring element

NOTES:

Chapter 2:
SKETCHER

*Tutorial 2.3: Advanced Profiles
& Sketch Analysis*

Featured Topics & Commands

Prerequisite Knowledge & Commands

- Entering workbenches
- Entering and exiting the *Sketcher* workbench
- Simple profiles
- Simple pads and pockets
- Applying dimensional and geometric constraints
- Work modes (*Sketch tools* toolbar)

Open and Closed Profiles

A profile is a series of geometric elements (points, lines, arcs, etc...) connected together. Profiles are used to create or modify solids. If they are not constructed properly, CATIA will not allow you to perform certain operations in the *Part Design* workbench. Most operations such as *Pad* and *Pocket*, with some exceptions, require that a profile be closed and not self intersecting. (Note: CATIA V5R16 will *Pocket* and *Pad* open profiles, however, this is not recommended.)

- Open profile: A profile is open when the first and last elements of the series are not connected.

- Closed profile: A profile is closed when the last element connects up with the first element in the series.

- Self intersecting profile: A self intersecting profile has at least two elements crossing each other.

- Inner profiles: Inner profiles are closed profiles inside other closed profiles. The operation

performed on an inner profile is opposite to the operation performed on the outer profile. For example, if the outer profile is padded then the inner profile is pocketed.

The Value Field

When sketching in the *Sketcher* workbench, a value field appears that allows you to enter exact coordinates, lengths, angles and radii. The value field attaches itself to the *Sketch tools* toolbar. Consider the use of the value field when the *Profile* command is active. For the first point of the profile, you can enter a horizontal and vertical coordinate in the value field. The 'Tab' key is used to access the H: and V: value fields. The 'Enter' key should not be used when entering values in the value field.

If you are drawing a line, the second point of the profile can be defined by a coordinate point (H,V) or a length and an angle (L,A).

For an arc, the second point is defined by a coordinate point (H,V) and the radius (R) of the arc may also be entered.

Each 2D geometry (Line, Rectangle, Elongated Hole, Arc, etc...) has its own specific value field that allows you to control the geometric characteristics of that geometry.

Sketch Status

The sketch status tells you if your sketch is under-constrained, over-constrained, or iso-constrained. Without constraints (geometrical and dimensional) a sketch can be moved freely just by clicking and dragging on an element. This could cause problems. If an element within the sketch is moved, it affects the solid that the sketch supports and any part that is related to the solid through assembly constraints. Therefore, it is important to completely constrain each sketch. If a sketch is completely constrained, it turns green.

- <u>Under-constrained:</u> A sketch is under-constrained when it is not completely constrained. In simple terms, there is not enough given information to duplicate the sketch exactly.

- <u>Over-constrained:</u> A sketch is over-constrained when there are duplicate dimensions given. The dimensions will turn purple when it is over-constrained. The duplicate dimension may be deleted or converted into a reference dimension.
- <u>Iso-constrained:</u> A sketch is iso-constrained when the sketch is completely constrained and there are enough dimensions and constraints given to duplicate the sketch exactly.
- <u>Inconsistent constraints:</u> Inconsistent constraints are constraints that conflict. For example, a parallel and perpendicular constraint applied to the same two lines is inconsistent or impossible. If there are inconsistent constraints present, the part will turn red.

Sketch analysis and sketch solving status

Within the *Tools* toolbar there is a *2D Analysis* toolbar containing commands that will help determine the sketch status and analyze sketch problems.

- <u>Sketch Solving Status icon:</u> This command will tell you if your sketch is under, over, or iso-constrained.

- <u>Sketch Analysis icon:</u> This command analyzes your sketch to determine problem areas. For example, it will identify any place where your sketch is open or intersecting, and it will identify duplicate and inconsistent constraints.

Part Modeled

The part modeled in this tutorial is shown on the right. The main or outer profile is created using the help of the value field. The profiles used to modify the main pad are constructed in such a way as to illustrate the usefulness of solving for sketch status and analyzing sketches.

Section 1: Using the value field.

1) Enter a **New Part** drawing and if asked, name your part **Wedge**.

2) Enter the **Sketcher** 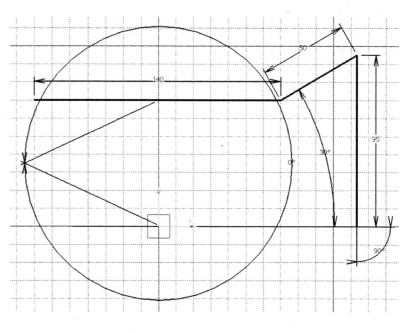 on the **yz plane**.

3) Enter your Options window (**Tools – Options...**). In the Sketcher section, activate the grid's Display and Snap to Point options, and set the snap spacing for both directions to be: Primary spacing = **100 mm,** Graduations = **10 mm**. In the Constraint section, activate both the Dimensional and Geometrical constraints.

4) Select the **Profile** icon.

5) Position your cursor at **-70,70** and click your left mouse button. Notice that a value field appears attached to your Sketch tools toolbar. Some of the value field may not be visible. If this is the case, move the toolbar so that all parts are visible.

End Point: H: 50mm   V: 10mm   L: 105.981mm   A: 233.324deg

6) Hit the **TAB** key until the L: field is activated. Type **140** in the L: field and hit the **TAB** key (do not hit the enter key).

7) Move your cursor around the screen. Notice that the angle of the line changes, but the length is constrained to be 140 mm and does not change.

8) Hit the **TAB** key until the A: field is activated, type **0**, and hit the **TAB** key. Notice that dimensions are automatically applied for every value that was typed in the value field.

(Having trouble? If your value field is not retaining the values that you type, move the cursor off the graphics screen and try again. If the cursor moves on the graphics screen, the values in the value field change to match the cursor position.)

9) Use the same procedure to make the next two lines. First line: L = **50** and A = **30**, Second line: L = **95** and A = **270**.

10) In the value field, click on the **Tangent arc** icon. Notice that the field options change.

Tangent Arc

11) Hit the **TAB** key until the R: field is activated. Type **60**, and hit the **TAB** key.

12) Move your cursor around the graphics screen. Notice that the end point of the arc changes but not the radius.

13) Move your mouse until a vertical constraint line appears and click for the endpoint of the arc. The arc should travel through 90°.

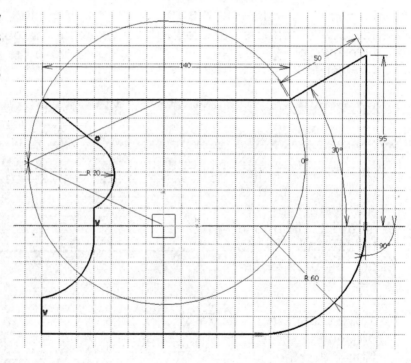

14) Using the value field, create two more lines: H = **-70**, V = **-60**, and then H = **-70**, V = **-40**.

15) In the value field, click on the **Three point arc** icon. Type the following values for the arcs second point: H = *-50*, V = *-30*.

Three point arc

16) Move your mouse around the graphics screen and notice that the radius of the arc changes.

17) Type the following values for the arcs third point: H = *-40*, V = *-10*.

18) Create a **Line** (select this in the value field) whose endpoint is H = *-40*, V = *10*.

19) Create a **Three point arc** whose second point is H = *-30*, V = *20* and whose radius is R = *20*.

20) Move your mouse around the graphics screen. Notice that radius does not change. Move the mouse until a vertical constraint with the last line drawn is created and click.

21) Snap to profile's start point.

(Notice that every value that was entered into the value field has a dimension associated with it.)

22) **Exit** the *Sketcher* and **Pad** the sketch to a length of *50 mm*.

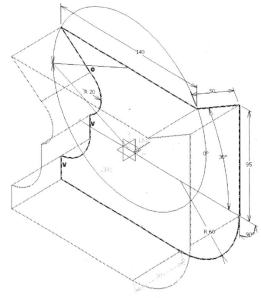

Section 2: Sketch Analysis

1) Enter the **Sketcher** 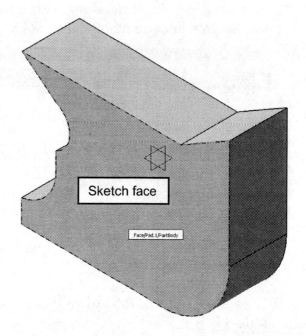 on the left face of the part.

Sketch face

Face/Pad.1/PartBody

2) Draw the **Profile** 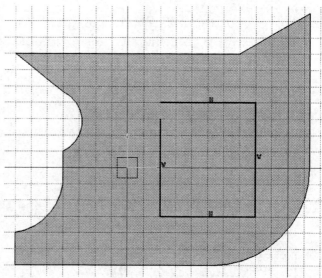 shown in the figure. Double click at the last point to end the *Profile* command.

3) Select the **Sketch Analysis** icon. This is located in the *Tools* toolbar and usually found in the bottom toolbar area. The *Sketch Analysis* icon is usually stacked under the *Solve Sketch Status* icon.

4) A *Sketch Analysis* window will appear. Select the **Geometry** tab. In the Detailed information area, it should say that your profile is open. Notice that on your sketch two blue circles appear that identify the open section.

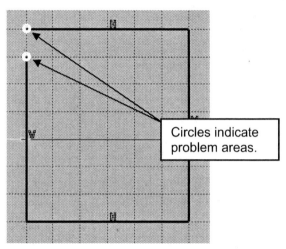

Circles indicate problem areas.

5) **Close** the *Sketch Analysis* window.

6) Draw a **Line** that starts at the end of the open vertical line and crosses the top horizontal line of the sketch.

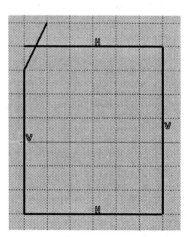

7) **Exit** the *Sketcher* and try to apply a **Pocket** to the sketch. An error window will appear. This is because the profile is self intersecting. Select **OK** in the error window and **Cancel** the *Pocket Definition* window.

8) Edit Sketch.2 (the most recent sketch). This may be done by double clicking on the sketch in the tree or on the part, or right clicking on Sketch.2 in the specification tree and selecting Sketch.2 object – Edit.

9) Select the **Sketch Analysis** icon.

10) A *Sketch Analysis* window will appear. Select the **Geometry** tab. In the Detailed information area, it should say that your profile is autocrossing or self intersecting. Notice that on your sketch two blue circles appear that identify the problem area.

11) **Close** the *Sketch Analysis* window.

12) Select the **H** constraint of the top horizontal line and hit the **Delete** key. Then drag the lines left endpoint up to coincide with the top endpoint of the angled line to create a closed profile.

13) Perform a **Sketch Analysis** 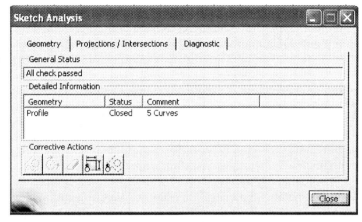 and notice that the sketch is now closed.

14) Draw a **Circle** inside the profile as shown in the figure.

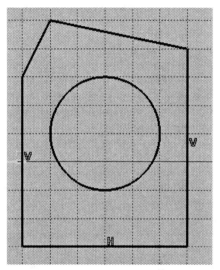

15) **Exit** the *Sketcher* and **Pocket** to the sketch to a depth of *20 mm*. Notice that the outer profile is pocketed and the inner profile is not.

16) Enter the **Sketcher** on the left face of the part.

Sketch face

17) On your own, do your best to create the **Elongated Hole** shown. Remember to read the prompt line for help. The *Elongated Hole* icon is stacked under the *Rectangle* icon.

18) **Exit** the *Sketcher* and **Pocket** to the sketch using the option of **Up to Last**.

Section 3: Solving for the sketch status.

1) Rotate the part so that you can see the back face of the part. You may do this by using the middle + right mouse buttons, or by using the compass.

2) Enter the **Sketcher** on the face indicated in the figure.

Sketch face

3) Activate your **Construction/Standard Element** icon.

4) Use the **3D Project Element** icon to project the edges indicated in the figure onto the sketch plane.

5) Deselect all.

6) Deactivate your **Construction/Standard Element** icon.

Projected edges

7) Draw the **Rectangle** shown making sure to make coincidence constraints with the top and bottom projected lines. You will know if a coincident constraint is applied if a double circle symbol appears. Notice that the top and bottom lines are green. They are constrained.

8) Select the **Sketch Solving Status** icon. A *Sketch Solving Status* window will appear and tell you that the sketch is under-constrained. **Close** the *Sketch Solving Status* window.

9) **Constrain** 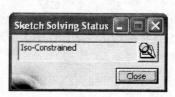 the rectangles width (**10 mm**) and distance from the side of the part (**20 mm**) as shown in the figure. Notice that once the constraints are applied, the rectangle turns green.

10) Select the **Sketch Solving Status** icon. A *Sketch Solving Status* window will appear and tell you that the sketch is iso-constrained (completely constrained). **Close** the *Sketch Solving Status* window.

11) **Constrain** 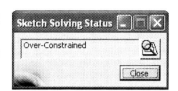 the rectangles height as shown in the figure. Notice that the dimension is purple indicating that it is a duplicate dimension.

12) Use the **Sketch Solving Status** icon to determine the sketch status.

13) Double click on the duplicate height dimension. In the *Constraint Definition* window activate the **Reference** toggle. Notice the parentheses now enclose the dimension value and the rectangle again becomes iso-constrained.

14) Apply a perpendicular constraint between the left and right vertical lines of the rectangle. Select the left and right vertical lines using the **CTRL** key. Select the **Constraint Definition** icon and activate the **Perpendicular** toggle. Notice that part of the rectangle turns pink or red indicating inconsistent constraints.

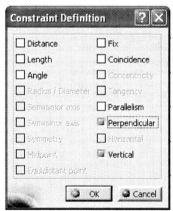

15) Use the **Sketch Solving Status** icon to determine the sketch status.

16) Select the green horizontal line between the left and right sides of the rectangle (the perpendicular constraint) and delete it. The rectangle should now be iso-constrained.

17) **Exit** the *Sketcher* and **Pad** the rectangle up to the curved surface. In the Pad Definition window; Type: **Up to Surface**, Limit: select the curved surface as shown in the figure. If you get an error window, select the **Reverse Direction** in the *Pad Definition* window.

Select this surface as the limit.

Chapter 2: SKETCHER

Tutorial 2.4: Modifying Geometries & Relimitations

Featured Topics & Commands

Prerequisite Knowledge & Commands

- Entering workbenches
- Entering and exiting the *Sketcher* workbench
- Work modes (*Sketch tools* toolbar)
- Creating profiles
- Applying dimensional and geometric constraints
- Editing a preexisting sketch
- Simple pads and pockets
- Rotating the part using the mouse or compass

Modifying geometries using the mouse

The mouse can be used to move different elements of a sketch just by clicking and dragging that element. The way in which the element moves depends on the constraints that are present. For example, a horizontally constrained line can be moved anywhere but it will always remain horizontal.

The Operation toolbar

The *Operation* toolbar contains commands that allow you to modify and transform elements of a sketch. The commands contained in the *Operations* toolbar from left to right are

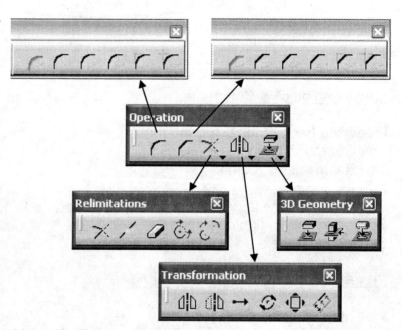

- Corner: The *Corner* or fillet command creates a rounded corner. When selected, five options appear in the *Sketch tools* toolbar. Each option is a different method of dealing with the corner after the fillet is applied. The options from left to right are: *Trim all Elements, Trim First Element, No Trim, Standard Line Trim, Construction Line Trim,* and *Construction Lines No Trim.*
- Chamfer: The *Chamfer* command creates an angled corner. When selected, five options appear in the *Sketch tools* toolbar. Each option is a different method of dealing with the corner after the chamfer is applied. The options from left to right are: *Trim all Elements, Trim First Element, No Trim, Standard Line Trim, Construction Line Trim* and *Construction Lines No Trim.*
- Relimitations toolbar: This toolbar contains commands that allow you to re-limit or modify 2D geometries. It contains commands such as *Trim, Break,* and *Close.*
- Transformation toolbar: This toolbar contains commands that allow you to duplicate and modify existing geometries. The commands from left to right are: *Mirror, Symmetry, Translate, Rotate, Scale,* and *Offset.*
- 3D Geometry toolbar: This toolbar contains commands that allow you to use 3D geometries to create 2D sketches. The commands from left to right are: *Project 3D Elements, Intersect 3D Elements,* and *Project 3D Silhouette Edges.*

The Relimitations toolbar

The *Relimitations* toolbar contains commands that allow you to re-limit or modify 2D geometries. The commands from left to right are

- Trim: The *Trim* command allows you to delete unwanted portions of an element. The *Trim* command allows you to dynamically choose how you want the element trimmed. When selected, two options appear in the *Sketch tools* toolbar: *Trim All Elements* and *Trim First Element*. If unfamiliar with this command, it may be a bit confusing.
- Break: The *Break* command allows you to break one element into two.
- Quick Trim: The *Quick Trim* command allows you to delete unwanted portions of an element. When selected, three options appear in the *Sketch tools* toolbar: *Break and Rubber In, Break and Rubber Out*, and *Break and Keep*. When the *Break and Rubber In* option is selected then the selected element will be automatically trimmed between the two closest limiting elements. When the *Break and Rubber Out* option is selected then the selected element will be kept and the elements on the outside of the two closest limiting elements will be automatically trimmed. When the *Break and Keep* option is selected the selected element will be broken at the two closest limiting elements.
- Close: The *Close* command creates a complete circle out of an arc. This command may also be used on ellipses.
- Complement: The *Complement* command erases an arc and creates its complement. This command may also be used on ellipses.

Part Modeled

Two different views of the part modeled in this tutorial is shown on the right. While constructing the first sketch, the mouse will be used to move and modify the sketch. Modifications to the main pad will be constructed using the commands located in the *Operation* and *Relimitations* toolbars.

Section 1: Modifying geometries with the mouse.

1) Open a **New... Part** drawing and, if asked, name it **Angled Pin**.

2) Enter the **Sketcher** on the **xy plane**.

3) Activate **Geometrical** and **Dimensional** Constraints (orange), and deactivate the **Snap to Point** (blue).

4) Draw a **Rectangle** similar to that shown in the figure. Since the geometrical constraints are active, horizontal and vertical constraints are automatically applied.

5) Deselect all.

6) Using your mouse, click and drag the right side vertical line to the left. Notice that the horizontal and vertical constraints are maintained, but the rectangle may move up and down because it is not constrained in space.

7) Click and drag the bottom right corner point up and to the left.

8) **Constrain** the left and top lines of the rectangle as shown. (Only select one line when constraining. Do not select 2 lines to place one dimension.)

9) Click and drag the rectangle to the center. Notice that the whole rectangle moves. The rectangle is completely constrained with respect to itself, but it is not constrained in space and therefore is not green.

10) **Constrain** the rectangle in space as shown. Notice that the rectangle is now iso-constrained and is green.

11) Try to move the rectangle with the mouse. You should not be able to.

12) **Exit** the *Sketcher*

and **Pad** the sketch to a length of **100 mm**.

13) Edit Sketch.1.

14) Delete the bottom horizontal (H) and right side vertical (V) constraints, and the constraints (dimensions) locating the rectangle in space.

15) Click and drag the bottom right corner point up and to the left.

16) **Exit** the *Sketcher*. Notice that the Pad.1 is automatically updated to reflect the changes made in the sketch.

Section 2: Re-limiting geometries

1) Deselect all.

2) Enter the **Sketcher** on the right face of the part.

3) Pull out the *Relimitations* toolbar.

Sketch face

Face/Pad.1/PartBody

4) Draw and **Constrain** the following profile using the following commands: **Circle**, **Bi-Tangent Line**. When completed, your sketch should be iso-constrained.

5) Trim the part of the circles that lie between the tangent lines. Double click on the

   Quick Trim icon and then select the parts of the circle that lie between the tangent lines as indicated in the figure. Hit the **ESC** key to exit the *Quick Trim* command.

50

25

D30

45

Trimmed elements

D30

50

(Note: If we tried to pad the sketch as is, CATIA would give us an error of self intersecting elements. Therefore, we need to break the tangent lines where they cross.)

6) Select the **Break** icon and then select the lines one after the other. Read the prompt line to understand why you are selecting the lines. Select the

   Break icon again. This time select the lines in reverse order.

7) **Exit** the *Sketcher* and

Pad the sketch to a length of *40 mm*.

8) Deselect all.

9) Enter the **Sketcher** on the face of one of the padded circles as indicated in the figure.

Sketch face

10) Use the **3D Project Elements** icon to project both arcs onto the sketch plane. Note that they are yellow and still associated with the part.

(Note: We are going to use the *Close* and *Complement* commands on these projected arcs. These commands can not be performed on associated (yellow) elements. Therefore, we need to disassociated or isolate the elements from the part.)

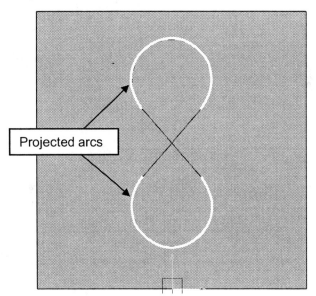

Projected arcs

11) Isolate the projected elements. Use the **CTRL** key and mouse to select both projected elements. At the top pull down menu, select **Insert – Operation – 3D Geometry – Isolate**. Notice that the elements are no longer yellow.

12) On your own, use the **Close** 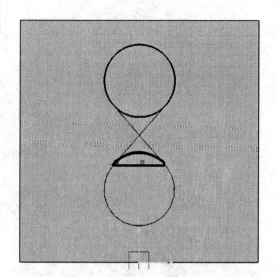 command to close the top arc and the

Complement command to create a complement of the bottom arc. Then,

draw a **Line** between the endpoints of the bottom arc.

13) **Exit** the *Sketcher* and **Pad** the sketch to a length of *10 mm*.

Section 3: Corners and Chamfers

1) Deselect all.

2) Rotate the part around and enter the

Sketcher on the back face of the part as indicated in the figure.

Sketch face

Face/Pad.1/PartBody

3) Draw and **Constrain** ⬚⊤ the two

Rectangles ▢ shown in the figure. Place the constraints on the sides shown.

4) Use **Quick Trim** ✐ to trim off the inside lines of the rectangle as shown.

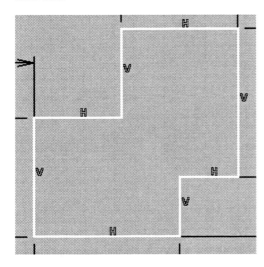

5) Apply *Corners* (fillets) to the inside corners of the sketch and change the dimension values to those shown in the figure. The corners should be applied with the trim all option activated. For the **8 mm** corner, select **Corner**

⌐ and then select the corner point. Pull the radius out and click. Change the dimension value by double clicking on the dimension. For

the **5 mm** corner, select **Corner** ⌐ and then select both lines that make the corner.

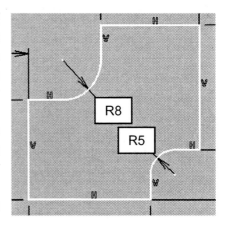

6) Try to apply a **Chamfer** to the upper right corner. An error window may appear basically telling you that you are about to apply inconsistent constraints. Get out of the error window and undo (**CTRL + Z**) the chamfer.

7) Delete the 25 mm and 20 mm height and width dimensions of the upper rectangle.

8) Reapply the 20 mm and 25 mm **Constraints** selecting the 2 limiting lines instead of the line length.

9) Apply a **Chamfer** to the upper right corner of the sketch of length *10 mm* and angle *45°*.

10) Use procedures similar, re-constrain and apply a **Chamfer** to the lower left corner of the sketch of length *10 mm* and angle *45°*.

11) **Exit** the Sketcher and **Pocket** ▣ the sketch using the option **Up to Last**.

NOTES:

Chapter 2: SKETCHER

Tutorial 2.5: Axes & Transformations

Featured Topics & Commands

Prerequisite Knowledge & Commands

- Entering workbenches.
- Entering and exiting the *Sketcher* workbench
- Work modes (*Sketch tools* toolbar)
- Creating profiles
- Applying dimensional and geometric constraints
- Editing a preexisting sketch
- Simple pads and pockets
- Rotating the part using the mouse or compass

Axes

An axis is a line of symmetry. The *Axis* command is located in the *Profile* toolbar. When an axis line is draw it is automatically drawn as a construction

element. Therefore, it does not become part of the final sketch. Axes are often used to facilitate the drawing process when using commands such as *Mirror*. It may also be used as the rotation axis when creating a *Shaft* (a revolved solid) or a *Groove* in the *Part Design* workbench.

The Transformation toolbar

Transformation tools allow you to change, or create an array of a sketch. These tools allow you to use an existing sketch to define new a sketch that is a mirror images, a rotated version, or a different size from the original sketch. Many of the transformation commands may be used in 'duplicate' mode which enables arrays to be created. The *Transformation* toolbar in located in the *Operation* toolbar. Reading from left to right, the *Transformation* toolbar contains the following commands

- Mirror: The *Mirror* command creates a mirror image of a sketch about a selected axis.
- Symmetry: The *Symmetry* command replaces a sketch with its mirror image.
- Translate: The *Translate* command moves a sketch from one point to another. This command may be used in duplicate mode to create a linear array.
- Rotate: The *Rotate* command rotates a sketch around a selected center point. This command may be used in duplicate mode to create a polar array.
- Scale: The *Scale* command scales a sketch relative to a selected point. This command may be used in duplicate mode which creates several instances of the sketch.
- Offset: The *Offset* command creates a sketch that is offset from the original by a specified distance. This command scales and adjusts the sketch as appropriate. Given the complex nature of this command, it does not always work as intended.

Part Modeled

The part modeled in this tutorial is shown on the right. The outside hub is created using an *Axis* and the *Shaft* command. The spokes are created using the *Rotate* command, and the triangular *Pockets* are created using the *Mirror* command.

Section 1: Using Axes

1) Open a **New... Part** drawing and, if asked, name it **Wheel**.

2) Enter the **Sketcher** 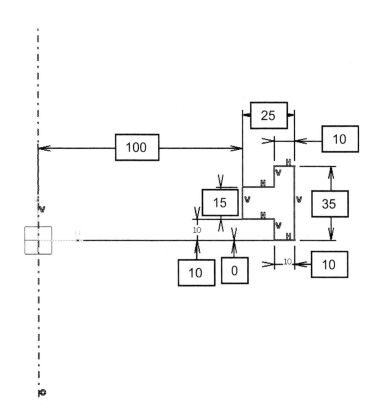 on the **yz plane**.

3) Draw an **Axis** that is on and aligned with the vertical (V) axis.

4) Draw and **Constrain** the **Profile** shown.

5) **Exit** 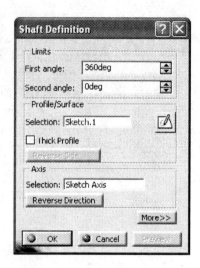 the Sketcher and **Shaft** the sketch using the Sketch Axis as the rotation axis. In the *Shaft Definition* window fill in the following fields; First angle: **360**, Second angle: **0**, Profile Selection: **Sketch.1**, and Axis Selection: **Sketch Axis**.

6) In the *Part Design* workbench apply fillets to the inside corners of the hub. Select the **Edge Fillet**

 icon. In the *Edge Fillet Definition* window, set the fillet radius to **2 mm**. Activate the *object(s) to fillet:* field and select all 6 inside edges and then select **OK**. (UNIX users will need to hold the CTRL key down when selecting the edges.)

Edges to fillet.
(Same edges on the other side.)

7) Deselect all and enter the **Sketcher** on the **yz plane**.

8) Select the **Cut Part by Sketch Plane** icon located in the *Tools* toolbar usually in the bottom toolbar area.

9) Draw an **Axis** that is on and aligned with the V axis.

10) Draw and **Constrain** the **Profile** shown.

11) **Exit** the Sketcher and **Shaft** the sketch **360** degrees using the Sketch Axis as the rotation axis.

Section 2: Transformations

1) Create a new sketch plane by offsetting the xy plane by 12 mm. If your *Reference Element (Extended)* toolbar is not active, activate it. At the top pull down menu select **View – Toolbars – Reference Elements (Extended)**. If it is not apparent where this toolbar ended up, look in the corners for any hidden toolbars. Select the **Plane**

icon. In the *Plane Definition* window fill in the following fields; Plane type: **Offset from plane**, Reference: right click and select **xy plane**, Offset: *12 mm*. Then select **OK**. A new plane should be created 12 mm from the xy plane.

2) Enter the **Sketcher** on the newly created plane.

3) Pull out the Transformation subtoolbar.

4) Draw and **Constrain** the following **Ellipse** . To constrain the major axis just click on the ellipse. To constrain the minor axis click on the ellipse, when the major axis dimension appears, right click and select **Semiminor axis**.

5) Array the ellipse to create a total of 8 ellipses each 45 degrees apart. Select the ellipse and then select the

Rotate icon. In the *Rotation Definition* window select **7** Instance(s), and activate the **Duplicate mode**. Notice that the Angle field does not allow you to enter a value yet. Look at the Prompt line. Select the center point to be the coordinate origin, enter an angle of **45deg** and then select **OK**. (UNIX users need to hold the CTRL key down when the center point is selected.)

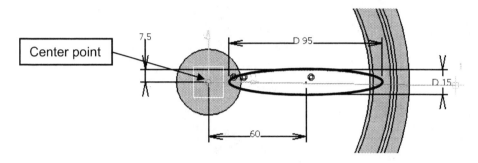

Center point

6) **Exit** the Sketcher and **Pad** the sketch to a length of **11 mm**.

7) Select the **Edge Fillet** icon. In the *Edge Fillet Definition* window, set the fillet radius to *2 mm*. With the object(s) to fillet: field active select all 16 side faces of the padded ellipsis and then select **OK**. (UNIX users will need to hold the CTRL key down when selecting the faces.)

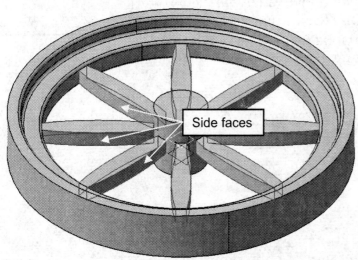

Side faces

8) Enter the **Sketcher** on the top face of the center hub.

Sketch face

9) Draw and **Constrain** the following **Axis** and **Profile**.

10) Use the *Symmetry* command to mirror the original sketch to the other side of the axis. Select the entire triangle and not the axis. Select all elements of the triangle, select the **Symmetry** 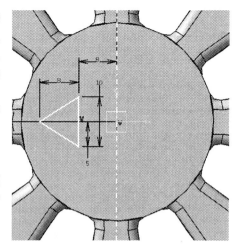 icon and then select the axis.

11) On your own, use the **Mirror** 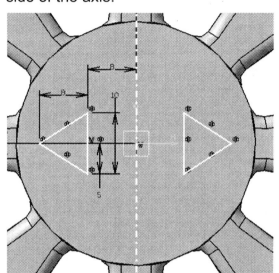 command to create a copy of the sketch on the other side of the axis.

12) **Rotate** both triangles through a distance of *90deg* creating *1* instance in the **Duplicate mode**.

13) **Exit** the Sketcher and **Pocket** the
sketch using the option **Up to last**.

Chapter 2: SKETCHER

Tutorial 2.6: Operations on 3D Geometries & Sketch planes

Featured Topics & Commands

Prerequisite Knowledge & Commands

- Entering workbenches
- Entering and exiting the *Sketcher* workbench
- Work modes (*Sketch tools* toolbar)
- Creating profiles
- Applying dimensional and geometric constraints
- Editing a preexisting sketch
- Simple pads and pockets
- Rotating the part using the mouse or compass

The 3D Geometry toolbar

Many times it is necessary to relate the current sketch to the existing part. There are two ways of doing this. You can constrain the sketch using the edges of the part as a point of reference, or you can project the part's edges and silhouettes onto the sketch plane. This second method of projection has many advantages. It is quick (you don't have to redraw the sketch) and the projection is associative. An associative projected sketch is connected to the existing part and automatically updates when the part is changed. The default color of an associative sketch is yellow. The *3D Geometry* toolbar is located in the *Operation* toolbar. The commands in the *3D Geometry* toolbar from left to right are

- Project 3D Elements: This command allows you to project an edge or surface of a part onto the sketch plane.
- Intersect 3D Elements: This command allows you to project the intersection between the part and the sketch plane.
- Project 3D Silhouette Edges: This command allows you to create a silhouette of the part. You can only create a silhouette edge from a canonical surface whose axis is parallel to the sketch plane.
- Isolate: *Isolate* is a command that is not located in the *3D Geometry* toolbar but is used in conjunction with the above three commands. *Isolate* breaks the link between the projected sketch and the part. Once the *Isolate* command has been applied to a sketch, the sketch will no longer be associative. To access this command select <u>I</u>nsert – <u>O</u>peration – <u>3</u>D Geometry – <u>I</u>solate.

The Reference Element toolbar

Thus far our sketch planes have been the 3 principle planes or a face of the part. Many times we need a sketch plane that can not be defined by original coordinate system or the part. The commands located in the *Reference Element* toolbar allows us to create planes and other reference elements that may be used as the sketch plane or help in the creation of a sketch plane. These elements are not standard elements and will not become part of your solid. The *Reference Element* toolbar is located in the *Part Design* workbench and is not normally active. To activate this toolbar select <u>V</u>iew – <u>T</u>oolbars – *Reference Elements (E<u>x</u>tended)*. The commands located in the *Reference Element* toolbar from left to right are

- <u>Point:</u> Allows you to create a reference point.
- <u>Line:</u> Allows you to create a reference line. Mainly used to define a direction for another command.
- <u>Plane:</u> Used to create a reference plane to be used as a sketch plane. There are many ways to create a reference plane. A couple examples include; a plane offset from an existing plane or part face, and a plane at an angle to another plane of part face.

Part Modeled

The part modeled in this tutorial is shown on the right. This part is constructed using projections and reference planes. Notice that several reference planes have been created to facilitate the drawing process.

Section 1: 3D Project Elements

1) Open a **New... Part** drawing and, if asked, name it **Angled Channel**.

2) Enter the **Sketcher** on the **yz plane**.

3) Draw and **Constrain** the following **Profile** and **Circle** shown. Notice that the circle's center is coincident with the origin.

4) **Exit** the *Sketcher* and **Pad** the sketch to a length of **20 mm**.

5) Create a reference plane that is offset from the yz plane by 100 mm. If the *Reference Element (Extended)* toolbar is not active, activate it. At the top pull down menu select **View** – **Toolbars** – **Reference Elements (Extended)**.

Select the **Plane** ◇ icon. In the *Plane Definition* window fill in the following fields; Plane type: **Offset from plane**, Reference: right click and select **yz plane**, Offset: *100 mm*.

Offset plane

6) Enter on the **Sketcher** on the newly created plane.

7) Select the **Isometric View** icon.

8) Display the **Grid**.

9) Use the **3D Project Elements** 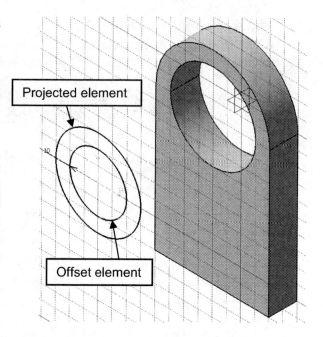 to project the outline of the hole onto the sketch plane. Notice that the projection is yellow. This means that it is associated with the part.

10) Offset the projection 10 mm to the inside. Select the **Offset** icon (located in the *Transformation* toolbar), select the projection, move the mouse to the inside of the circle and click, and then double click on the dimension and change its value to *10 mm*.

11) Draw a horizontal **Line** between quadrants of the outside projected circle.

12) Use the **Quick Trim** command to trim off the top half of the two circles as well as the portion of the line inside the smaller circle. The trimmed portions of the associated elements will turn into construction elements.

13) **Exit** the *Sketcher* and **Pad** the sketch to a length of ***180 mm*** in the **Reverse Direction**.

14) Edit Sketch.1.

15) Change the hole diameter to ***40 mm***.

16) **Exit** the *Sketcher*. Notice that Pad.2 which is based on the projection updates automatically.

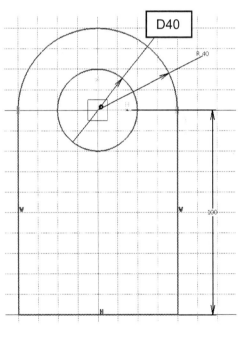

17) Edit Sketch.2.

18) Use a selection window to select all the elements in Sketch.2. Then at the top pull down menu, select **Insert – Operation – 3D Geometry – Isolate**. Notice that the projections are no longer yellow. They are not associated with the part anymore.

19) **Exit** the *Sketcher* and edit Sketch.1.

20) Change the hole diameter back to **60 mm**.

21) **Exit** the *Sketcher*. Notice that Pad.2 does not update this time.

Section 2: Project 3D Silhouette Edges

1) Offset a **Plane** **10 mm** from the **yz plane**.

2) Enter on the **Sketcher** on the newly created plane.

3) Draw and constrain the following Sketch.

4) **Exit** the *Sketcher* and **Shaft** the sketch through **360deg** using the **V Direction** as the rotation axis.

5) Offset a **Plane** *100 mm* from the **zx plane**.

6) Enter on the **Sketcher** on the newly created plane.

7) Select the **Isometric View** icon.

8) Try using **3D Project Elements** to project the Shafted sphere. You can't do it. All you get is a line.

9) Use **Project 3D Silhouette Edges** to project the silhouette of the two protruding parts of the sphere.

10) Draw **Lines** to connect the top ends and bottoms ends of the projected arcs.

11) Delete the parallel constraint between the two lines if one appears.

12)**Exit** the *Sketcher* and **Pad** the sketch to a length of *100 mm* in the **Reverse Direction**.

13)Edit Sketch.3. This is the sketch associated with the *Shaft*.

14)Change the radius dimension to *25 mm*.

15)**Exit** the *Sketcher*. Notice that Pad.3 automatically updates.

Section 3: Intersect 3D Elements.

1) Create two reference points. Select the **Point**

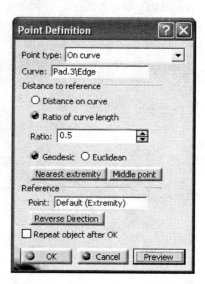

icon. In the *Point Definition* window fill in the following fields; Point type: **On curve**, Curve: select the top line shown in the figure, Distance to reference: activate **Ratio of curve length**, Ratio: *0.5*. Create a second point on the bottom line in a similar fashion.

Select as curve for 1st point.

Points

2) Create a reference **Line**

between the two points.

Point 1

3) Select the **Plane**
icon. In the *Plane Definition* window fill in the following fields; Plane type: **Angle/Normal to plane**, Rotation axis: click on the **Line.1** (the reference line) Reference: select the surface indicated in the figure, Angle: *30deg*.

Reference surface

4) Enter on the **Sketcher** on the newly created plane.

5) Select the **Isometric View** icon.

6) Select the **Intersect 3D Elements** icon. Then select Pad.3 (the last pad) in the specification tree.

16) **Exit** the *Sketcher* and **Pad** the sketch to a length of **50 mm** in the **Reverse Direction**.

NOTES:

Chapter 2: SKETCHER

Tutorial 2.7: Points & Splines

Featured Topics & Commands

Prerequisite Knowledge & Commands

- Entering workbenches
- Entering and exiting the *Sketcher* workbench
- Work modes (*Sketch tools* toolbar)
- Creating profiles
- Applying dimensional and geometric constraints
- Editing a preexisting sketch
- Simple pads and pockets
- Rotating the part using the mouse or compass

Points

Points are used to define some aspects of 2D geometries. For instance, a line is defined by two end points, and a circle is defined by a center point and a radius. When creating geometries such as lines, circles, rectangles and profiles, points are automatically created by CATIA. However, points may also be created by the user. User created points are used most often when creating splines and wireframes. The *Point* toolbar is located within the *Profile* toolbar. Reading from left to right, the *Point* toolbar contains the following commands

- Point by Clicking: Allows you to create a point by clicking anywhere on the screen. The position of this point may be changed later by double clicking on and entering a new position in the *Point Definition* window.
- Point by Using Coordinates: Allows you to create a point by entering a Cartesian or polar coordinate in the *Point Definition* window.
- Equidistant Points: Allows you to create a specified number of equidistant points along a previously drawn curve.
- Intersection Point: Allows you to create a point at the intersection of two elements.
- Projection Point: Projects (in 2D) a point onto a pre-existing profile. The projection line which connects the point with the profile creates a 90° angle with the profile.

Splines

Splines are used when a profile is curved and complex. Such profiles are too complex to be created using arcs and conics. Reading from left to right, the *Spline* toolbar contains the following commands.

- Spline: A spline is a curved profile defined by three or more points. The tangency and curvature radius at each point may be specified.
- Connect: The connect command connects two points or profiles with a spline.

Part Modeled

The part modeled in this tutorial is a cam. The cam profile is created by using construction points and a spline. The cam profile is then modified by entering the *Spline Definition* window and adding and modifying the points used to create the spline. The cam is shown at two different stages of the tutorial. The original cam profile is shown below. The cam profile graph shows the follower displacement versus the rotation angle. The cam has a base circle radius of 1.5 inches.

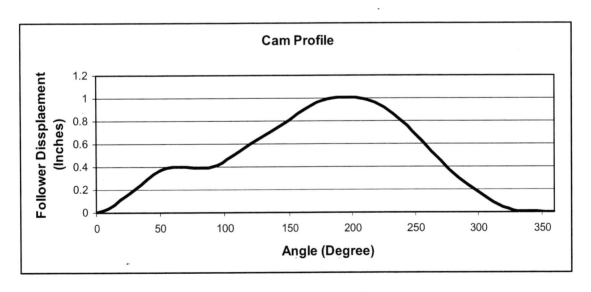

Section 1: Entering points.

1) Open a **New... Part** drawing, name it *Cam* and enter the **Sketcher** 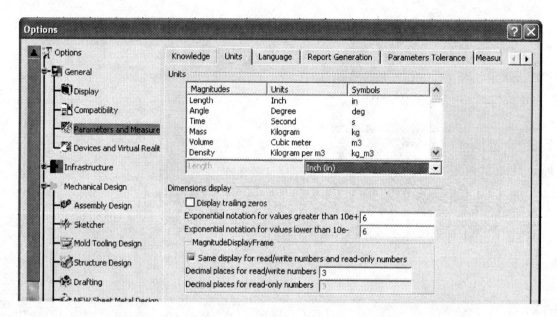 on the **yz plane**.

2) At the top pull down menu, select **Tools – Options...** Deactivate the grid display and snap to point options. In the left side menu of the *Options* window, select **Parameters and Measure**. Click on the **Units** tab and set the length units to **Inches**.

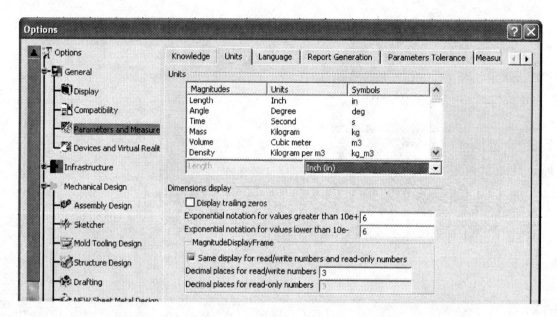

3) Activate the **Construction/Standard Element** icon. We want the points that we will be creating to be construction elements.

4) <u>Double click</u>* on the **Point by Using Coordinate** icon. In the *Point Definition* window click on the **Polar** tab and enter a radius of *1.5in* and an Angle of *0deg*. Then select **OK**. When the next *Point Definition* window appears, click on the **Polar** tab and enter a radius of *1.7in* and an Angle of *30deg*. Continue filling in the *Point Definition* window with the following Cam profile data. Remember to add the base circle radius to each displacement point. We have already entered the first two points. Start with the third.

* Double clicking allows you to continually enter points that are relative to the origin. If you don't double click you will end up entering points that are relative to the last point created unless you deselect the last point each time.

Angle (Degrees)	Follower Displacement (Inches)
0	0
30	0.2
60	0.4
90	0.4
105	0.49
120	0.6
135	0.705
150	0.81
165	0.92
180	1
210	1
225	0.92
240	0.79
255	0.62
270	0.45
285	0.3
300	0.18
315	0.07
330	0.01

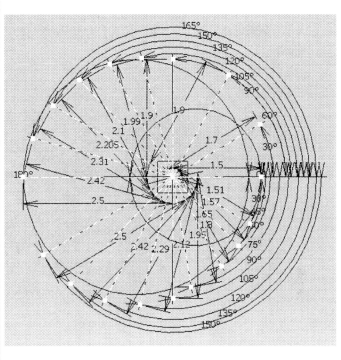

Section 2: Creating and editing splines.

1) Deactivate the **Construction/Standard Element** icon.

2) Click on the **Spline** icon. Move your mouse to the first created point (1.5in<0deg). When a bull's-eye symbol appears click. Select each successive point until you return to the first point. Double click on the first point to end the spline.

Bull's-eye symbol

3) **Exit** the *Sketcher* and **Pad** the sketch to a length of **.25in** in both directions (**Mirrored extent**).

4) Enter the **Sketcher** on the **yz plane**.

5) Draw a **Circle** that is centered at the origin and has a diameter of **1.25 in**.

6) **Exit** the *Sketcher* and **Pad** the sketch to a length of **1in** in both directions (**Mirrored extent**).

7) Enter the **Sketcher** 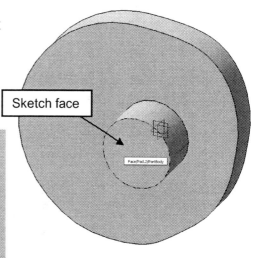 on the front face of the center hub.

8) Draw and constrain the following sketch.

Sketch face

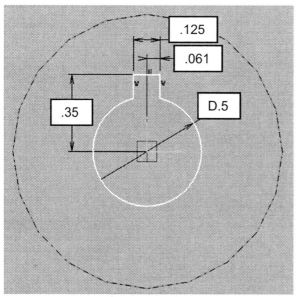

9) **Exit** the *Sketcher* and **Pocket** the sketch using the option **Up to last**.

10) Edit Sketch.1 and double click anywhere on the spline. In the *Spline Definition* window, click on **CtrPoint.4** and activate **Replace Point**. Click on the graphics screen somewhere above CtrPoint.4 making sure not to create any constraints. Click **OK**.

11) Double click on the newly created point. In the *Point Definition* window, click on the **Polar** tab and enter a radius of *2.1in* and an angle of *90deg*.

12) Double click on the spline. In the *Spline Definition* window click on **CtrlPoint.3** and activate **Add Point After**. Click somewhere between CtrlPoint.3 and CtrlPoint.4 making sure not to create any constraints.

13) Edit the new point's position and enter R = **2in** and A = **80deg**.

14) Double click on the spline. In the *Spline Definition* window click on **CtrlPoint.21** the point between CtrlPoint.3 and CtrlPoint.4. Activate both the **Tangency** and **Curvature Radius** toggles. Enter a Curvature Radius of **.4in**. Look at the spline while you click **OK** to see the effect that changing the curvature radius has on the spline.

15) **Exit** the *Sketcher*.

16) Change the diameter of inner hub from 1.25 inches to **1 inch**.

NOTES:

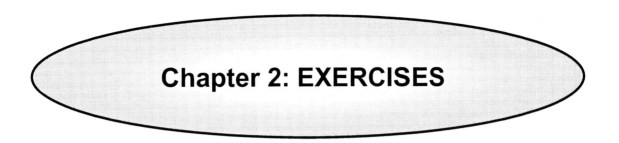

Exercise 2.1: This exercise can be performed after completing tutorial 2.1. Model the following parts using the *Snap to Point* command to estimate distances. You will be creating a 3-D models based on the 2-D orthographic projections shown.

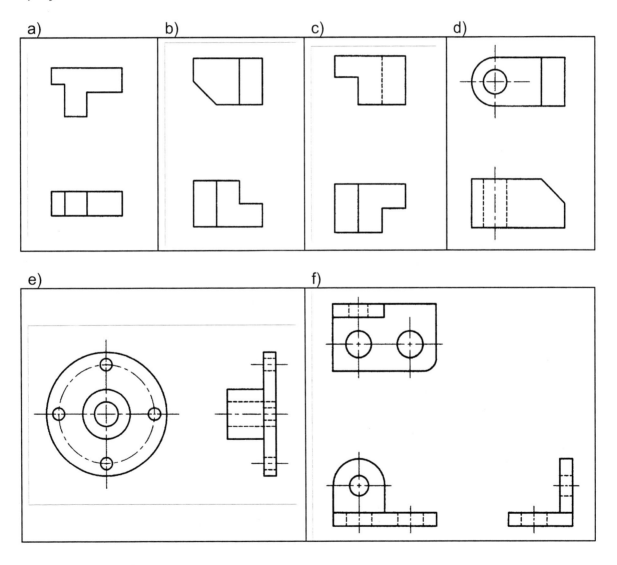

Exercise 2.2: This exercise can be performed after completing tutorial 2.2. Model the following parts applying the appropriate constraints. To set the appropriate units, select _Tools – Options... – Parameters and Measure_.

a)

b)

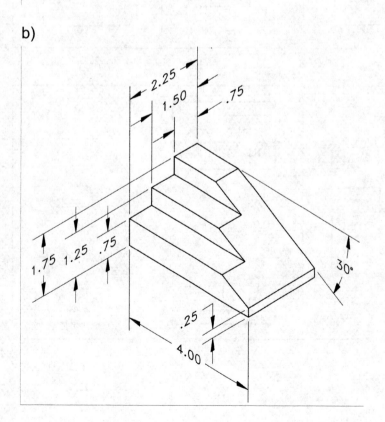

Exercise 2.3: This exercise can be performed after completing tutorial 2.3. Model the following parts applying the appropriate constraints. To set the appropriate units, select _Tools_ – _Options..._ – _Parameters and Measure_.

a)

b)

c)

d)

Exercise 2.4: This exercise can be performed after completing tutorial 2.4. Sketch the following parts applying the appropriate constraints. To set the appropriate units, select *Tools – Options... – Parameters and Measure*.

a)

b)

Exercise 2.5: This exercise can be performed after completing tutorial 2.5. Model the following parts applying the appropriate constraints, and transformations during the sketch phase. Use at least one transformation command to create the object. To set the appropriate units, select *Tools – Options... – Parameters and Measure*.

a)

b)

Exercise 2.6: This exercise can be performed after completing tutorial 2.6. Model the following parts applying the appropriate constraints, and using reference elements and projections where necessary. To set the appropriate units, select _Tools_ – _Options... – Parameters and Measure_.

a)

SECTION B-B

SECTION A-A

b)

Chapter 3: PART DESIGN

Introduction

Chapter 3 focuses on CATIA's *Part Design* workbench. The reader will learn how to use sketches to create solids and how to create solids without the aid of sketches.

Tutorials Contained in Chapter 3

- Tutorial 3.1: Sketch-Based Features: Pads
- Tutorial 3.2: Sketch-Based Features: Pockets, Holes & Slots
- Tutorial 3.3: Sketch-Based Features: Shafts & Grooves
- Tutorial 3.4: Sketch-Based Features: Advanced commands
- Tutorial 3.5: Dress-Up Features: Drafts, Fillets & Chamfers
- Tutorial 3.6: Dress-Up Features: Shell, Thickness & Threads
- Tutorial 3.7: Transformation Features
- Tutorial 3.8: Boolean Operations
- Tutorial 3.9: Material Properties

NOTES:

Chapter 3:
PART DESIGN

Tutorial 3.1: Sketch-Based Features: Pads

Featured Topics & Commands

Prerequisite Knowledge & Commands

- The *Sketcher* workbench and associated commands
- Editing a preexisting sketch
- Simple pads and pockets
- Rotating the part using the mouse or compass

The Part Design workbench

The *Part Design* workbench is used to create and modify solids. A solid is first created using an initial sketch. Once the main solid is created, it may be modified using sketch based features or using commands that do not require a sketch. The general design process is to; sketch the profile of the main pad, generate the main pad, create additional sketch based features, add dress-up features, modify and reorder features as desired, and insert new bodies for more complex parts.

The Sketch-Based Features toolbar

The commands located in the *Sketch-Based Features* toolbar use sketches to create or modify a solid. Therefore, to modify the shape of the feature, the sketch must first be modified. The depth, length of travel, rotation angle, etc... may be modified in the respective definition windows. It is possible to modify a feature (Pad, Pocket, etc...) after it has been created. Just double click on the respective feature in the specification tree and the *Definition* window will appear allowing you to change the parameters. Reading left to right, the commands in the *Sketch-Based Features* toolbar are

- Pad: *Pads* are used to add or create solid material in a linear direction.
- Pocket: *Pockets* are used to remove material in a linear direction.
- Shaft: *Shafts* are used to add or create material by revolving a sketch about an axis.
- Groove: *Grooves* are used to remove material by revolving a sketch about an axis.
- Hole: The *Hole* command is used to create holes of different configurations. For example, this command could be used to create a simple blind hole or a more complex counter bore and drilled hole, or even a threaded hole.
- Rib: The *Rib* command is used to add material by sweeping a profile along a center curve.
- Slot: *Slots* are used to remove material by sweeping a profile along a center curve.
- Advanced Sketch-Based Features toolbar: This toolbar contains the commands, *Solid Combine* and *Stiffener*. *Stiffener* is used to create relatively thin supporting structures for your part.
- Loft: *Lofts* are used to create or add material by connecting several sketches.
- Remove Loft: *Remove Loft* is used to remove material by connecting several sketches.

The Pads toolbar

The *Pads* toolbar, along with the simple *Pad* command, contains advanced commands for creating pads. The *Pads* toolbar reading from left to right contains the following commands.

- Pad: This command is used to create a simple pad.
- Drafted Filleted Pad: This command is used to create a pad that has angled sides and rounded edges.
- Multi-Pad: This command is used to create pads of varying depth all within a single sketch.

Part Modeled

The part modeled in this tutorial is shown on the right. The entire part is created exclusively using the commands located in the *Pads* toolbar and their supportive sketches.

(Important! It is not uncommon for CATIA to tell you that your sketch is over-constrained. If this occurs, look to the coincident constraint. These are usually the culprit. CATIA may create a coincident constraint between two distance corners. Delete this constraint and try again.)

Section 1: Creating Multi-Pads

1) Open a new part drawing and enter the **Sketcher** on the **xy plane**. Deactivate the Grid's Snap to Point and Display.

2) Before you start to draw make sure your length units are set to millimeters. (**Tools – Options... – Parameters and Measures**)

3) Draw and **Constrain** the following **Rectangle**.

4) **Exit** the *Sketcher.*

5) Pull out the *Pads* toolbar [image].

6) **Pad** [icon] the sketch to a length of **20 mm**. Select **OK**. In the specification tree, double click on Pad.1. The Pad Definition window will appear. If needed you could change the pad features. Select **Cancel**.

7) Enter the **Sketcher** on the top face of the block.

8) **3D Project Elements** the face of the block on to the sketch plane as a **Construction Element**.

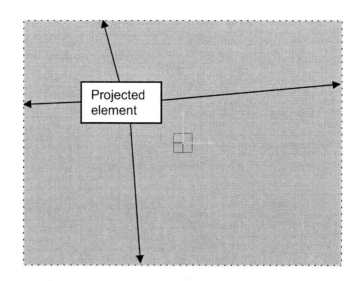

9) Draw and **Constrain** the following **Rectangles**. Make sure to snap exactly to the corner.

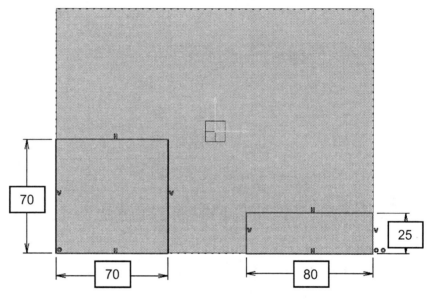

10) Add the following **Profiles** to the sketch. Make sure not to draw overlapping lines. CATIA may make an unwanted coincident constraint that will make the sketch over constrained.

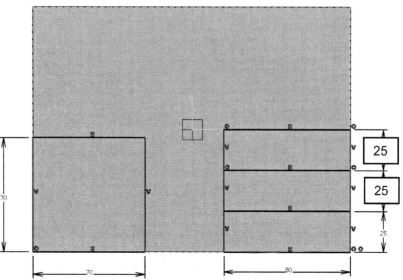

11) **Exit** the Sketcher and select the **Multi-Pad** icon. In the *Multi-Pad Definition* window click on one of the Extrusion domains in the Domains field. Notice that a feature will highlight in blue on the part. Type in that feature's length and click on the next, etc... The pad lengths are as follows. The lone rectangle is *5 mm*, the corner rectangle in the group of three is *10 mm*, the next is *20 mm*, and the last is *30 mm*.

Sketch face

Section 2: Advanced Pads

2-1: Adding and removing elements of a sketch.

CATIA allows you to *Pad* only a portion of a sketch. The user may manually add or remove elements from the profile definition.

1) Enter the **Sketcher** on the top face of the main pad as shown in the figure above.

2) **3D Project Elements** the top face of the main pad on to the sketch plane as a **Construction Element**.

3) Draw the two profiles shown in the figure.

4) **Exit** the Sketcher and deselect all.

5) Select the **Pad** icon. In the *Pad Definition* window, right click on the Selection: field and click on **Go to profile definition**. In the *Profile Definition* window, activate the **Sub-element** toggle and click the **Add** button and select the profile indicated in the figure. Select **OK**. In the *Pad Definition* window, set the length to *30 mm* and apply the pad.

Add this profile.

2-2: Padding using an open profile.

CATIA allows you to *Pad* an open profile as long as it is bounded by the part.

1) Enter the **Sketcher** on the top face of the main pad.

2) Right click on **Sketch.3** (the last sketch) in the specification tree and select **Hide/Show**.

Sketch face

3) **3D Project Elements** the top face of the main pad on to the sketch plane as a **Construction Element**.

4) Draw and **Constrain** the closed **Profile** shown in the figure.

5) **Exit** the *Sketcher* and **Pad** the sketch to a length of **30 mm**.

6) Enter the **Sketcher** on the top face of the main pad.

Sketch face

7) **3D Project Elements** the top face of the main pad on to the sketch plane as a **Construction Element**.

8) Draw a **Spline** similar to that shown in the figure.

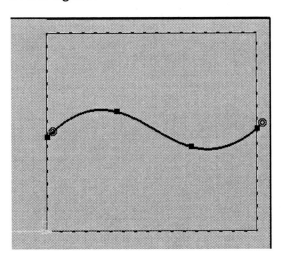

9) **Exit** the *Sketcher* and select the **Pad** icon. Set the length to **20 mm** and select the **Preview** button. Select the **Reverse Side** button and **Preview** again (you may have to click on the screen to reactivate the preview button). Notice the difference. Apply the *Pad* when it looks like the figure.

2-3: Padding to a surface.

CATIA allows you to *Pad* up to a surface or offset from the surface. If an offset option is used, the end of the *Pad* takes on the same shape as the surface.

1) Rotate the part and enter the **Sketcher** on the face indicated in the figure.

Sketch face

2) **3D Project Elements** the sketch face on to the sketch plane as a **Construction Element**.

3) Draw and **Constrain** the following **Profile**. Note that the centers of the two arcs are coincident (the same). This constraint may be applied after the two arcs are drawn.

4) **Exit** the *Sketcher* and **Pad** the sketch to a length of **120 mm**.

Sketch face

5) Enter the **Sketcher** on the top flat face of the part.

6) **Cut Part By Sketch Plane**
.

7) **3D Project Elements** the sketch face on to the sketch plane as a **Construction Element**.

8) Draw and **Constrain** the **Rectangle** shown in the figure. To place the 10 mm dimension you may have to hide the projected construction element.

9) **Exit** the Sketcher and **Pad** the sketch using the option of **Up to surface**. Activate the Limit: field and select the surface indicated in the figure and set the Offset: to *-10 mm*.

2-4: Padding with thickness.

CATIA gives you the option of padding a sketch with thickness. In simple terms, it means that you can easily create a thin walled section.

1) Rotate the part and enter the **Sketcher** on the face indicated in the figure.

Sketch face

Face/Pad.2/PartBody

2) **3D Project Elements** the left edge of the sketch face on to the sketch plane as a **Construction Element**.

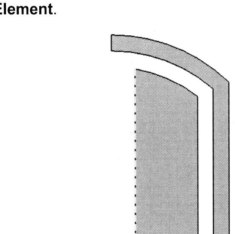

3) Draw and **Constrain** the **Rectangle** shown in the figure.

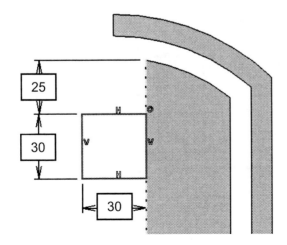

25

30

30

4) **Exit** the Sketcher and select the

Pad icon. Set the length to be **120 mm** and select the **More>>** button. On the left side of the window, activate the **Thick** toggle. On the right side of the window enter a value of **2 mm** for Thickness 1.

Sketch face

Section 3: Creating Drafted Filleted Pads

1) Enter the **Sketcher** on the right face of the part as shown in the figure above.

2) Draw and **Constrain** the **Rectangle** shown in the figure.

3) Exit the Sketcher and select the **Drafted**

Filleted Pad icon. Set the Length: to be **50 mm**. Activate the Limit: Field and select the sketch face. The draft angle is **5deg** and all the fillets are **5 mm**.

NOTES:

Chapter 3:
PART DESIGN

Tutorial 3.2: Sketch-Based Features: Pockets, Holes & Slots

Featured Topics & Commands

Prerequisite Knowledge & Commands

- The *Sketcher* workbench and associated commands
- Editing a preexisting sketch
- Pads
- Rotating the part using the mouse or compass

The Pockets toolbar

The *Pockets* toolbar, along with the simple *Pocket* command, contains advanced commands for creating pockets. The *Pockets* toolbar, reading from left to right, contains the following commands.

- Pocket: This command is used to create a simple pocket.
- Drafted Filleted Pocket: This command is used to create a pocket that has angled sides and rounded edges.
- Multi-Pocket: This command is used to create pockets of varying depth all within a single sketch.

Holes

The *Hole* command is used to create holes of different configurations. For example, this command could be used to create a simple blind hole or a more complex counter bore and drilled hole, or even a threaded hole. The *Hole* command is initiated in the *Part Design* workbench without a preexisting sketch. However, the command creates a sketch; therefore, it is still a sketch based feature.

Slots

Slots are used to remove material by sweeping a profile along a center curve. The profile may be opened or closed depending on the options chosen. The center curve should not be made of multiple geometric elements.

Part Modeled

The part modeled in this tutorial is shown on the right. It is created from one main pad and then modified using the *Pocket, Hole* and *Slot* commands. Several different *Pocket* and *Hole* options are explored.

Section 1: Creating Pockets

1) Open a **New... Part** and if prompted to, name it **Pockets**.

2) Enter the **Sketcher** on the **xy plane**.

3) Draw and **Constrain** the following **Rectangle**.

4) **Exit** the *Sketcher* and **Pad** the sketch to a length of **80 mm**.

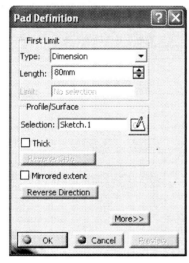

5) Enter the **Sketcher** on the top face of the block.

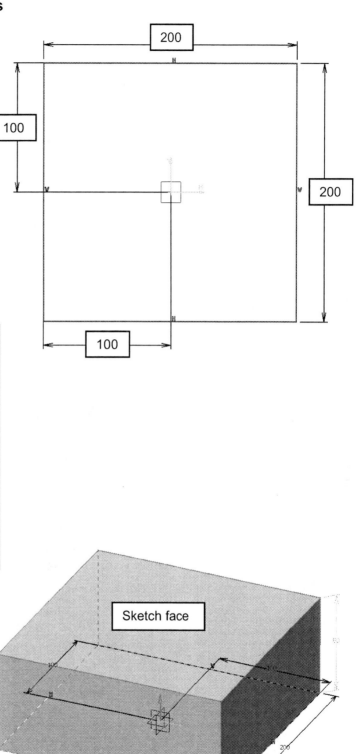

Sketch face

6) Draw and **Constrain** the following sketch. All the corner (fillet) radii are *5 mm* and all the hole diameters are *15 mm*. To avoid unwanted constrains, hold the **SHIFT** key down when drawing the circles.

7) Pull out the **Pockets** toolbar

8) **Exit** the *Sketcher* and select the **Multi-Pocket** icon. Set the depth of the inside holes to *80 mm* and the depth of the outside profile to be *10 mm*. Remember when you click on one of the Extrusion domains it will highlight the respective feature on the part.

Sketch face

9) Enter the **Sketcher** on the left face of the part as shown in the figure.

10) **3D Project Elements** the sketch face on to the sketch plane as a **Construction Element**.

11) Select the projection and then select the **Offset** icon. Click your mouse to the inside of the projection. Set the offset dimension to be *-15 mm*. Change the offset profile from a construction element to a **Standard Element**.

12) **Exit** the *Sketcher* and select the **Pocket** icon. In the *Pocket Definition* window select the **More>>** button. Then activate the **Thick** toggle. Set the pocket depth to be *20 mm* and both thickness' to be *3 mm*. Set and preview one thickness at a time to see witch direction is effected by both.

13) Enter the **Sketcher** on the left face of the part.

14) **3D Project Elements** the inside portion of the sketch face on to the sketch plane as a **Construction Element**.

15) Draw a **Spline** similar to the one shown in the figure.

16) **Exit** the *Sketcher* and **Pocket** ⊡ the sketch to a depth of **20 mm**. Use the *Reverse Side* button if necessary to achieve the same results as shown in the figure.

Sketch face

Section 2: Creating Slots.

1) Enter the **Sketcher** on the right face of the part as shown above.

2) **3D Project Elements** the top edge of the sketch face on to the sketch plane as a **Construction Element**.

3) Draw and **Constrain** the following **Profile**. The profile is a closed triangle.

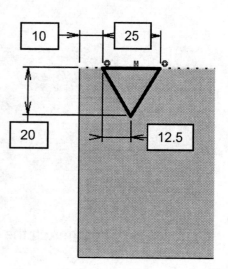

4) **Exit** the Sketcher, deselect all and re-enter the **Sketcher** on the top face of the part.

5) **3D Project Elements** the sketch face on to the sketch plane as a **Construction Element**.

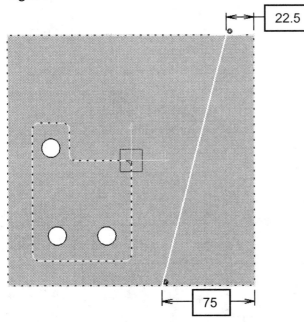

6) Draw and **Constrain** a **Standard Element Line** as shown in the figure.

22.5

75

7) **Exit** the Sketcher and deselect all.

8) Select the **Slot** icon. Use the triangle as the *Profile* and the line as the *Center curve*.

<u>Section 3:</u> Creating Holes.

1) Select the **Hole** icon and then select the top face of the part. Select the **Extension** tab in the *Hole Definition* window. Create a **Blind** hole with a diameter of *10 mm*, a depth of *10 mm*, and a **V-Bottom**. Click on the **Positioning Sketch** icon and **Constrain** the center of the hole as shown in the figure.

2) **Exit** the Sketcher and select **OK**.

3) Create another **20 mm** diameter through (**up to last**) **Hole** on the top face of the part. The hole should be constrained as shown in the figure.

4) Create a counterbored **Hole** with the following parameters. The counterbore diameter is **30 mm** with a depth of **10 mm**. The drill (hole) diameter is **20 mm** and it is a through hole. Select the **Type** tab to change the hole type from simple to **Counterbored**.

5) Create a countersunk **Hole** with the following parameters. The countersunk a depth of *10 mm* and an angle of *90deg*. The drill (hole) diameter is *20 mm* and it is a through hole.

6) Rotate your part so that the back side is visible.

7) Create a reference line that will control the direction of a hole. Select the **Line** icon. In the *Line Definition* window, fill in the following fields; Line type: **Angle/Normal to curve**, Curve: select the bottom edge of the back face, Support: select the back face, Point: select the bottom left corner of the face, Angle: *-45deg*, Length Type: **Infinite**.

Line Definition

Line type : Angle/Normal to curve

Curve: Hole.4\Edge

Support: Hole.4\Face

Point: Hole.4\Vertex

Angle: -45deg

Start: 0mm

End: 20mm

Length Type
- ○ Length   ○ Infinite Start Point
- ● Infinite   ○ Infinite End Point

☐ Mirrored extent

☐ Geometry on support

Normal to Curve

Reverse Direction

☐ Repeat object after OK

OK   Cancel   Preview

Support

Curve
(The bottom
edge)

Point
(The corner)

8) Create a *10 deg* **Tapered Hole** on the top face with a *10 mm* diameter that uses the reference line as its direction. To select the direction, deactivate the **Normal to surface** toggle, activate the field below it and select the reference line.

Chapter 3: PART DESIGN

Tutorial 3.3: Sketch-Based Features: Shafts & Grooves

Featured Topics & Commands

Prerequisite Knowledge & Commands

- The *Sketcher* workbench and associated commands
- Editing a preexisting sketch
- Pads
- Rotating the part using the mouse or compass

Shafts

A *Shaft* creates a solid by revolving a profile around an axis. Open or closed profiles may be used if the profile starts and ends at the axis. Only closed profiles can be used if the profile is away from the axis. The shaft can not be created if the profile crosses the axis. See the examples shown below.

- Closed or open section may be used when profile starts and ends on the axis.

- A closed profile must be used if the sketch does not contact the axis.

- Profiles that will not produce a shaft.

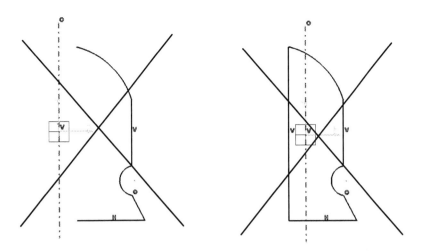

Grooves

A *Groove* removes material by revolving a profile around an axis. The profile may be open or closed depending on the options chosen in the *Groove Definition* window. But, the profile should never cross the axis.

Part Modeled

The part modeled in this tutorial is shown on the right. The main solid is created using a *Shaft*. Details are added using both the *Shaft* and *Groove* commands.

Section 1: Creating Shafts.

1) Open a **New... part** drawing, name it **Shaft** and enter the **Sketcher** on the **yz plane**.

2) Set your length units to be inches (**Tools – Options... – Parameters and Measures**)

3) Draw and **Constrain** the following **Profile** and **Axis**.

4) **Exit** the Sketcher and **Shaft** the sketch through an angle of **360deg** using the **Sketch Axis** as the Axis Selection.

5) Enter the **Sketcher** on the **zx plane**.

6) Use the **3D Project Element** and **Project 3D Silhouette Edges** to project all the edges of the part as **Construction Elements**.

7) Draw an **Axis** that is on and parallel to the V axis.

8) Using the projection as a guide, draw a **Profile** around the outline of the part on the right side of the axis as a **Standard Element**.

9) Offset the profile to the outside of the part. Select all elements of the profile

you created in step 8. Select the **Offset** icon. Activate the Offset: field in the Value field using the **TAB** key. Enter *.5* and hit the **TAB** key again.

10) Turn the original profile into a **Construction Element**.

11) **Exit** the Sketcher.

12) Select the **Shaft** icon. In the *Shaft Definition* window, set the First angle: to be *180deg*, **Reverse Direction** and use the **Sketch Axis** as the Axis Selection:. Select the More>> button and set Thichness1: to be *.125 in*.

13) Enter the **Sketcher** on the **zx plane** and draw and **Constrain** the **Rectangle** shown.

14) **Exit** the *Sketcher* and **Shaft** the sketch *180deg* using the Z-axis as the rotation axis.

Section 2: Creating Grooves.

1) Enter the **Sketcher** on the **zx plane**.

2) Draw and **Constrain** the following **Profile**.

3) **Exit** the Sketcher.

4) Select the **Groove** 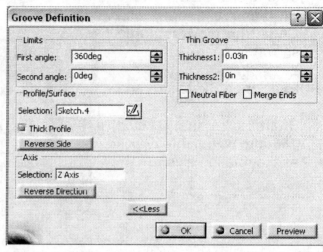 icon. In the *Groove Definition* window, set the First angle: to be **360deg** and use the Z-axis as the rotation axis. Select the **More>>** button and set Thickness1: to be *.03 in*. Select **OK**.

5) Double click on **Groove.1** in the specification tree.

6) In the Groove Definition window, set Thickness2: to be *.03 in*.

7) Enter the **Sketcher** on the **zx plane**. Draw and **Constrain** the **Circle** shown.

8) **Exit** the *Sketcher* and create a ***360 deg* Groove** using the sketch.

9) Edit **Shaft.3**. In the Shaft Definition window set the rotation angle to be **360deg**.

10) Enter the **Sketcher** on the **yz plane**.

11) Draw and **Constrain** the **Circle** shown. The center of the circle lies on the center axis.

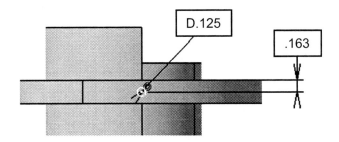

12)**Exit** the *Sketcher* and **Pad** the circle **2** inches in both directions.

Chapter 3: PART DESIGN

Tutorial 3.4: Sketch-Based Features: Advanced commands

<u>Featured Topics & Commands</u>

<u>Prerequisite Knowledge & Commands</u>

- The *Sketcher* workbench and associated commands
- Editing a preexisting sketch
- Pads
- Rotating the part using the mouse or compass

Lofts (Multi-sections Solid)

A *Loft* is generated by two or more planar sections that are swept along a spine. A *Loft* can add material or remove material. The planar sections may also be connected with guidelines which act as boundaries in which the *Loft* is contained. Direction arrows show the orientation of the *Loft* and can be used to change the orientation. *Lofts* are very useful in creating transitions between two existing solids.

Ribs and Slots

A *Rib* is a profile swept along a center curve (open or closed) to create a 3D feature. *Ribs* are useful when you need to sweep a profile from one surface to another. A *Rib* may also be used to create a pipe like feature by sweeping a profile along a center curve.

A *Slot* is a profile swept along a center curve (open or closed) to remove material from a solid.

The center curve used to create *Ribs* and *Slots* do not have to be drawn to the feature's true length. Walls of existing parts may be used as boundaries.

Stiffener

A *Stiffener* is a brace that is added to a wall or other structures that need to be strengthened to prevent possible breakage. It is often used in castings and molded plastics.

Part Modeled

The part created in this tutorial is a representation of a boat. Notice the sloping surface of the boats hull (loft), the thin support members (stiffeners), and the ribs and slots that are part of the outer shell.

Section 1: Loft Creation

1.1 – Creating a Simple Loft

1) Open a **New…** **Part** drawing, name it *Boat* and enter the **Sketcher** on the **zx-plane.**

2) Set your length units to millimeters.

3) Draw and **Constrain** the following **Ellipse**. The ellipses center coincides with the origin and the axes are along the V and H coordinate directions.

4) Draw a **Line** that cuts the ellipse in half as shown.

5) **Quick Trim** the top portion of the ellipse.

6) **Exit** the Sketcher and deselect all.

7) Create a new **Plane** by offsetting the **zx-plane** by **50 mm**. In the *Plane Definition* window, activate the **Repeat object after OK** toggle. After you select **OK** an *Object Repetition* window will appear. Enter **4** instances.

8) Create the same number of new planes on the other side of the sketch as shown.

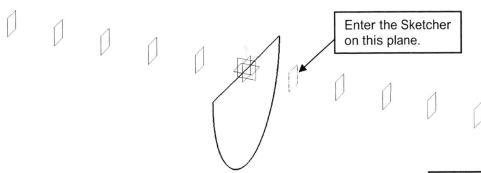

Enter the Sketcher on this plane.

Center point

9) Enter the **Sketcher** on the first new plane to the right of the sketch. See above figure.

10) Project Sketch.1 onto the new plane using **3D Project Elements**.

11) Scale the projection by 80% using the following procedure. Select all elements of the projection (Use the control key to select multiple elements). Select the **Scale** 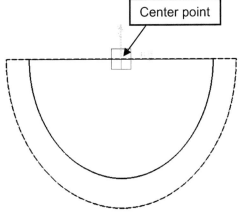 icon. Select the scaling center point as shown in the figure, activate the **Duplicate mode**, and enter **0.8** in the Value: field.

12) Turn the original projection into a **Construction Element**.

13)**Exit** the *Sketcher*.

14)Repeat the above procedure to create 4 more sketches all scaled 80% from that of the previously sketch.

15)Repeat for the other side.

16) Loft the 6 sketches on the right using the following procedure. Deselect all elements.

Select the **Loft** icon. Select the middle (biggest) sketch and then each sketch in order from left to right. After you have selected the last sketch on the right, select **OK**.

17) Repeat for the other side. However, the largest sketch now must be selected in the specification tree (this is because it is already part of a solid). The largest sketch which should be **Sketch.1**. The other sketches may still be selected manually with the mouse.

18) Pull out the *Dress-Up Features* toolbar .

19) **Shell** the solid to a thickness of **5 mm** removing the top face.

Face to remove

20) **Edge Fillet** the inside and outside joining line between the 2 halves of the boat with a radius of **20 mm**.

<u>1.2 – Relimiting a Loft</u>

By default a loft will begin and end with the planar section chosen. However, if you use guidelines or a spline to direct your loft, you can relimit the loft to begin or end with these curves.

1) Enter the **Sketcher** on the right side end of the boat.

2) **3D Project Elements** the end of the boat onto the sketch plane as a **Standard Element**.

3) **Exit** the *Sketcher.*

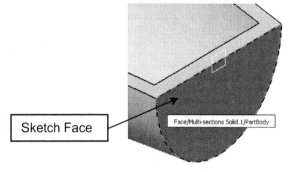

Sketch Face

4) Offset a **Plane** from the right end of the boat by *30 mm*.

5) Enter the **Sketcher** on the newly offset plane.

6) **3D Project Elements** the end of the boat onto the sketch plane as a **Construction Element**.

7) Scale the projection by a value of *0.1* using the same procedure as before.

8) Turn the scaled projection into a **Standard Element**.

9) **Exit** the *Sketcher* and re-enter the **Sketcher** on the **yz-plane**.

10) **3D Project Elements** the end of the boat onto the sketch plane as a **Construction Element**.

11) Draw and **Constrain** the following **Spline**.

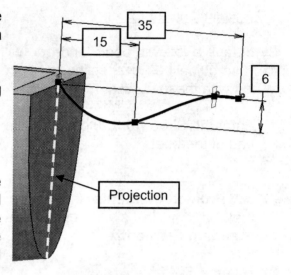

12) **Exit** the *Sketcher* and **Loft** 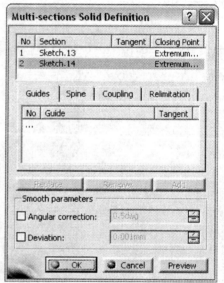 the end of the boat with the scaled version of the boat end. Select the projected sketch on the end of the boat and then the small scaled sketch.

13) If the loft looks funny, like shown below, edit *Multi-sections Solid.3* and change the direction of *Close point 2* by clicking on the arrow.

14) Double click on **Multi-sections Solid.3**. In the *Multi-sections Solid Definition* window, select the **Spline** tab. Activate the Spline: field and select the spline. Select **OK**. Notice the difference.

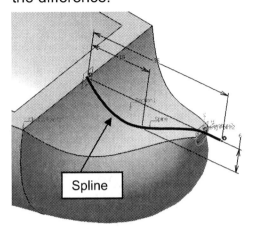

Spline

15) Double click on **Multi-sections Solid.3**. In the *Multi-sections Solid Definition* window, select the **Relimitation** tab and deselect the **Relimited on end section** toggle. Select **OK**. Notice the difference.

16) Double click on **Multi-sections Solid.3**. Remove the spline and reactivate the **Relimited on end section** toggle.

Section 2: Creating Ribs and Slots

2.1 – Creating a Simple Slot

1) Enter the **Sketcher** on the **xy-plane**.

2) **3D Project Elements** the outer profile of the boat onto the sketch plane. (The projection in the figure is shown black, however the real projection will be yellow.)

3) **Exit** the *Sketcher* and re-enter the **Sketcher** on the **zx-plane**.

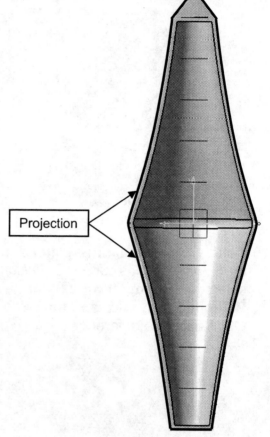

Projection

4) Draw and **Constrain** the following **Profile**. Use **Cut Part By Sketch Plane** to get a better view.

5) **Exit** the Sketcher and create a slot that goes around the boat using the above sketch and the

projection. Select the **Slot** icon. The Profile is the sketch created in step 4, and the Center curve is the projection created in step 3. If a *Warning* window appears, **Close** it.

6) Enter the **Sketcher** on the **xy-plane**.

7) Draw and **Constrain** the *5 mm* diameter **Circle** as shown.

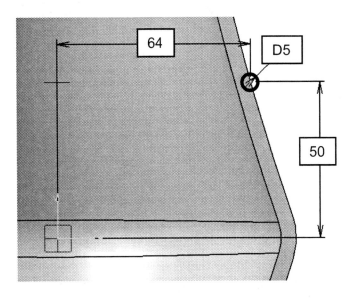

8) **Exit** the *Sketcher* and re-enter the **Sketcher** on the plane indicated in the figure. (The first offset plane created in this tutorial.)

Enter the *Sketcher* on this plane

9) **3D Project Elements** Sketch.2 (the sketch used to create the section of the boat) onto the sketch plane. Use **Cut Part By Sketch Plane** to get a better look.

Delete this line.

Projection

10) Isolate the projection and delete the top line of the sketch. (**Insert – Operation – 3D Geometry – Isolate**.)

11) **Exit** the *Sketcher* and create a rib using the above circle and projection. Select the **Rib** icon. Use the circle as the *Profile* and the projection as the *Center curve*.

12) Create a similar *Rib* on the other side as shown in the figure. This time you will project Sketch.7.

13) Enter the **Sketcher** on the **yz-plane**.

14) **3D Project Elements** the top edge of the left end of the boat on to the sketch plane. The projection will appear as a point.

Project this line

Projection point

15) Draw and **Constrain** the following **Profile**. Notice that the bottom of the profile is **Coincident** with the projection point and the top vertical line is **Tangent** with the radius.

16) **Exit** the *Sketcher* and re-enter the **Sketcher** on the top surface of the boat as shown in the figure.

Enter Sketcher on this surface

Face/Rib.1/PartBody

17) Draw and **Constrain** the *4 mm* diameter **Circle** shown.

18) **Exit** the *Sketcher* and create a **Rib** using the circle as the *Profile* and the vertical sketch as the *Center curve*.

19) Enter the **Sketcher** on the **yz-plane**.

20) Draw and **Constrain** the following **Rectangle**.

21) **Exit** the *Sketcher* and **Pad** the rectangle to a length of **3 mm** and **Mirror** the **extents**.

22) Enter the **Sketcher** on the front face of the rectangle (flag).

23)**3D Project Elements** the front face of the flag onto the sketch plane as a **Construction Element**.

24)Draw a **Spline** that starts and ends at the top of the flag like that shown in the figure. (Do not make it too complex or you may get an error when trying to apply the *Slot*.)

25)**Exit** the *Sketcher* re-enter the **Sketcher** on the top face of the flag.

26)Draw a **Circle** that is approximately center around the start point of your spline as shown in the figure.

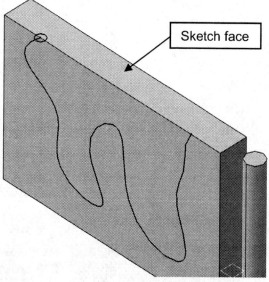

Sketch face

27)**Exit** the Sketcher and create a **Slot** using the circle as the Profile and the spline as the Center curve.

28)Create a **Slot** using the circle as the *Profile* and the spline as the *Center curve*. Activate the **Merge slot's ends** toggle. If a *Warning* window appears, **Close** it.

Section 3: Creating Stiffeners

1) Enter the **Sketcher** on the **xy-plane**.

2) **3D Project Elements** the left and right inner profiles of the boat on to the sketch plane as **Construction Elements**.

3) Draw and **Constrain** the **Line** shown in the figure.

Project these edges

200

4) **Exit** the *Sketcher* and create a stiffener that travels from the top down using the following procedure. Deselect all. Select the **Stiffener** icon. Activate the **From Top** toggle. Set Thickness1 to be **2 mm**. Activate the Selection: field and select the line.

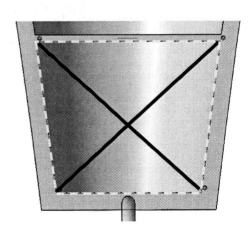

5) Enter the **Sketcher** on the **xy-plane**.

6) **3D Project Elements** the inner outline of the boat behind the stiffener as a **Construction Element**.

7) Draw the **Lines** shown.

8) **Exit** the *Sketcher* and create a **2 mm** thick **Stiffener** that travels **from** the **top** down using the crossed lines as the profile.

9) Enter the **Sketcher** on the offset plane (Plane.4) indicated in the figure.

10) **Cut Part By Sketch Plane** .

Sketch plane

11) **3D Project Elements** the profile of the boat at that section on to the sketch plane as **Construction Elements.** Use the projection to draw the 2 **Lines** shown. When creating the lines, error on the side of the curved projection.

12) **Exit** the *Sketcher* and create a **2 mm Stiffener**
 that travels **from** the **side** using the lines as
the profile.

13) On your own, draw and **Pad** a seat in the boat and support the seat with
a **Stiffener** as shown.

NOTES:

Chapter 3:
PART DESIGN

Tutorial 3.5: Dress-Up Features: Drafts, Fillets & Chamfers

Featured Topics & Commands

Prerequisite Knowledge & Commands

- The *Sketcher* workbench and associated commands
- Editing a preexisting sketch
- Pads and Pockets
- Rotating the part using the mouse or compass

The Dress-up Features toolbar

Dress-up features allow you to add finishing touches to your part. These commands are not based on a sketch. All parameters of the feature are controlled from the *Definition* window. The commands included in the *Dress-up Features* toolbar are

- The Fillets toolbar: The commands contained in the *Fillets* toolbar put radii on sharp corners. Several options are available.
- Chamfer: A *Chamfer* is a bevel created between 2 surfaces.
- The Drafts toolbar: The *Drafts* toolbar contains commands that will place an angle on a surface. Several options are available.
- Shell: The *Shell* command takes a solid part and hollows it out while keeping a given thickness on its sides.
- Thickness: The *Thickness* command adds material uniformly to a surface.
- Thread: Allows you to create threads or taps. The helix thread form is not shown in 3D; however, the information is stored and will show on a 2D drawing.
- The Remove Face toolbar: Allows you to remove or replace a face.

The Drafts toolbar

A *Draft* is an angled face. Material gets added or removed from your part when a draft is applied. Reading from left to right, the commands in the *Drafts* toolbar are

- Draft Angle: Applies a simple draft angle to a surface.
- Draft Reflect Line: This command allows you to draft from a line to a limiting element. It is a complex command and needs to be experimented with to understand its complete usefulness.
- Variable Angle Draft: This command allows you to specify different draft angles to a surface.

When applying a draft, you should keep in mind the following basic draft definitions.

- Pull direction: This is the direction that the draft angle is measured from.
- Draft angle: This is the angle that the draft faces make with the pulling direction.
- Neutral element: This is the element that remains unchanged after the draft command is applied.

- Parting element: This element defines where the draft is to start.
- Limiting elements: Define a region in which the draft should be applied.
- Points: This is used to define the locations where the draft angle varies when using a variable draft.

(Note: A draft angle can never be negative.)

The Fillets toolbar

A *Fillet* is a rounded edge or corner of constant or variable radius that is tangent to the two surfaces that create the edge. Reading from left to right, the commands in the *Fillets* toolbar are

- Edge fillet: This command creates a smooth transition between two surfaces.
- Variable radius fillet: This command allows the fillet to have more than one radius.
- Face-face fillet: This command is used when there is no intersection between the two surfaces or when there are more than two sharp edges between the surfaces.
- Tritangent fillet: This command involves removing one of the 3 surfaces selected.

Part Modeled

The part created in this tutorial is shown at two different stages of the process. Notice the angled surfaces (drafts), the rounded corners (fillets) and the beveled edges (chamfers).

Section 1: Creating Drafts

1.1 - Creating Simple Drafts

1) Enter a **New.. Part** drawing and name it **_Angled base_**.

2) Enter the **Sketcher** on the **yz-plane**.

3) Draw and **Constrain** the following **Profile**.

4) **Exit** the *Sketcher* and **Pad** the sketch to a length of **_100 mm_**.

5) Apply a constant **20º Draft Angle** to the right face of the part using the top face as the neutral element.

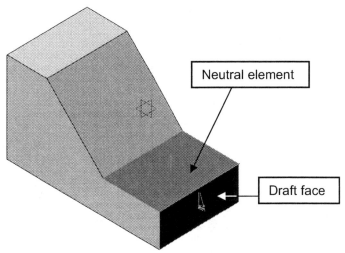

Neutral element

Draft face

Draft Definition

Draft Type:
Angle : 20deg
Face(s) to draft: 1 Face
☐ Selection by neutral face
Neutral Element
Selection: Pad.1\Face.2
Propagation: None
Pulling Direction
Pulling Direction Pad.1\Face.1
☑ Controlled by reference
More>>
OK   Cancel   Preview

6) Apply a constant **15º Draft Angle** to the front, back and left side faces of the part using all of the top faces as the neutral elements.

Draft Definition

Draft Type:
Angle : 15deg
Face(s) to draft: 3 Faces
☐ Selection by neutral face
Neutral Element
Selection: 3 elements
Propagation: None
Pulling Direction
Pulling Direction Pad.1\Face.2
☑ Controlled by reference
More>>
OK   Cancel   Preview

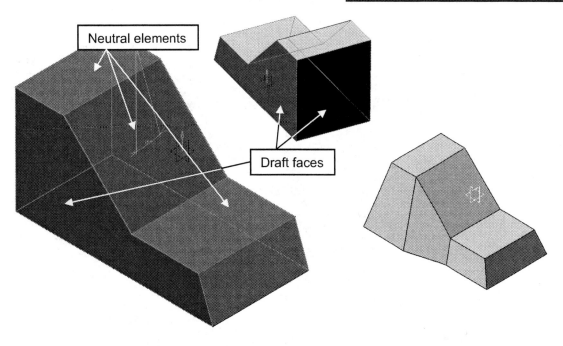

Neutral elements

Draft faces

7) Edit **Draft.2** (the draft applied to the 3 sides) by double clicking on it in the specification tree.

8) Change the neutral faces from the 3 top surfaces to just the bottom 1. Activate the *Neutral Element Selection* and click on the very top surface and the angled surface to deselect them.

(Notice: The neutral face indicated in the figure is the only face that remains unchanged.)

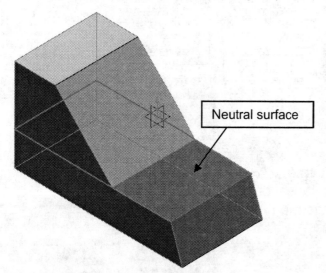

Neutral surface

1.2 - Using a Parting Element:

A *parting element* is a plane, face or surface that cuts the part in two. It is used as a boundary to tell the computer where to stop or change the draft angle.

1) Enter the **Sketcher** on the top face of the part.

2) Use **3D Project Element** to project the top face rectangle onto the sketch plane as a **Standard Element**.

Sketch face

Projected face

3) **Exit** the *Sketcher* and **Pad** 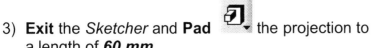 the projection to a length of **60 mm**.

4) Create a **Plane** that is offset **-20 mm** from the top surface of the part.

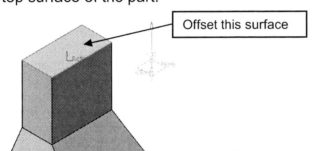

Offset this surface

5) Create a **10° Draft Angle** 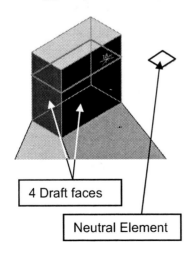 using the offset plane as a parting element. Select the 4 sides of the *Pad* as the Draft Faces and the offset plane as the Neutral Element. Select the **More>>** icon and activate the **Parting = Neutral** toggle. (Note: A curved surface may also be used as a parting element.)

4 Draft faces

Neutral Element

1.3 - Trimming Draft Features

1) Edit **Draft.1** (The first draft created on the right side face.) by double clicking on it.

2) Select the **More>>** icon in the *Draft Definition* window. Activate the Limiting Element(s): right click and select **Create Plane**. Create a plane that is offset **30 mm** from the **yz plane**. Repeat and create a plane that is offset **70 mm** from the **yz plane**. Click on the Orange arrows attached to the offset plane to change their direction to point inward towards each other and apply the draft.

3) Reenter the Draft Definition window and change the direction of the Orange arrows to the outward position and apply the draft and notice the difference.

1.4 - Variable Drafts:

A *Variable Angle Draft* allows you to change the draft angle along the surface.

1) Enter the **Sketcher** on the top of the base.

Sketch face

2) Draw and **Constrain** the **Rectangle** as shown. The rectangle is drawn **10 mm** away from all sides.

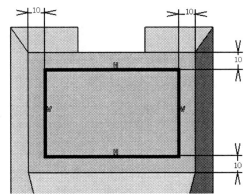

3) **Exit** the *Sketcher* and **Pocket** the sketch using the **Up to Last** option.

4) Create a **Variable Angle Draft** on the inside left face of the pocket using the top face as the Neutral Element. Once you pick a neutral element, the computer automatically selects 2 points. Add 2 more points by creating 2 planes that are offset from the **yz plane** by *30* and *70 mm*. Double click on the dimensions to change their values.

Section 2: Creating Fillets

2.1 - Creating Simple Fillets

1) Create a *5 mm* radius **Edge Fillet** on the top right side edge between the two drafted surfaces.

2) Create a **5 mm** radius **Edge Fillet** around the very top face of the part.

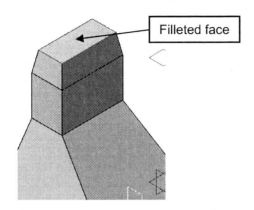

Filleted face

3) Enter the **Sketcher** on the very top surface of the part.

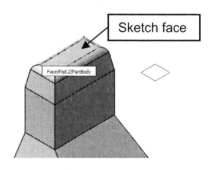

Sketch face

Face/Pad.2/PartBody

4) Draw and **Constrain** the **10 mm** diameter **Circle** shown. The circle is in the middle of the top surface.

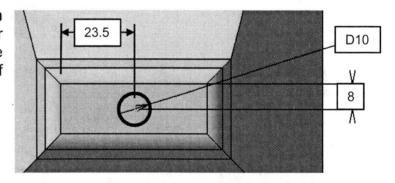

23.5

D10

8

5) **Exit** the *Sketcher* and **Pad** 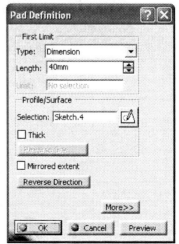 the circle to a length of **40 mm**.

6) Create a **5 mm** **Edge Fillet** around the bottom edge of the pin.

Filleted edge

2.2 - Rolling on an Edge:

The *Rolling on an Edge* option allows you to create a fillet that has a larger radius than the distance of the sides supporting the fillet.

1) Enter the **Sketcher** on the right face of the object between the drafted surfaces.

Sketch face

2) **3D Project Elements** the sketch face as a **Construction Element**.

3) Draw and **Constrain** the **Rectangle** as shown. Picking the top and bottom lines of the rectangle to create the 20 mm constraint will make it less likely that the rectangle will move out of position.

4) **Exit** the *Sketcher* and **Pad** the rectangle to a length of **40 mm**.

5) Enter the **Sketcher** on the same face as before.

6) **3D Project Elements** the sketch face as a **Construction Element**.

7) Draw and **Constrain** the **Rectangle** as shown.

8) **Exit** the *Sketcher* and **Pad** the rectangle to a length of **10 mm**.

9) Try to apply a **40 mm** radius **Edge Fillet** to the step. You will get an error because the step height is less than 40 mm. Select **OK** in the *Update Error* window. Select **Yes** in the *Feature Definition Error* window. When the *Edge Fillet Definition* window reappears, activate the *Edge(s) to Keep:* field and select the top edge of the step.

Filleted edge

Edge to keep

Filleted Edge

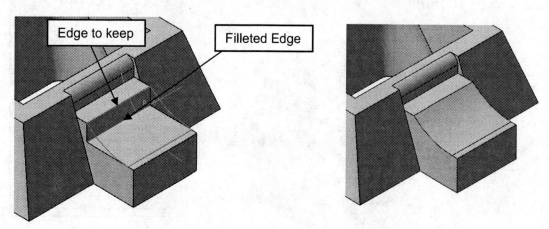

10) Enter the **Sketcher** on the **yz-plane**.

11) Draw and **Constrain** the **Rectangle** as shown. The reason that there are to vertical 30 mm constraints is that one constrains the height of the rectangle and one constrains the rectangle to the object.

12) **Exit** the Sketcher and **Pad** the rectangle using 2 limits. The first limit has a length of **120 mm** and the second has a length of **20 mm**.

13) Apply a **10 mm** **Edge Fillet** to the edges shown. (Notice that the edges between the filleted edges remain sharp.)

14) Reenter the *Edge Fillet Definition* window by double clicking on **EdgeFillet.5**. Select the **More>>** button. Keep the 2 edges indicated. (Notice that the edges are rounded now.) Repeat for the other side.

2.3 - Limiting Fillets:

Limiting a fillet allows you to apply a fillet to only a portion of an edge.

1) Apply a **10 mm** radius **Edge Fillet** to the top portion of the angled edges of the part using the **zx-plane** as a Limiting Element. Click on the orange arrow to change direction of fillet travel if necessary.

2.4 - Face to Face Fillets:

A *Face to Face Fillet* allows you to create a fillet between two surfaces that do not touch.

1) Delete **Draft.4** (The drafted surface inside the pocket) by right clicking on it in the specification tree and selecting **Delete**.

2) Delete the **Pocket.1**.

3) Enter the Sketcher on the top face of the base.

4) Draw and **Constrain** the *30 mm* diameter **Circles** shown.

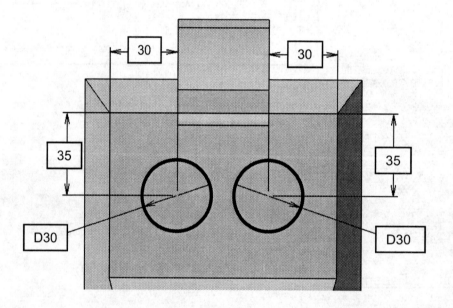

5) **Exit** the *Sketcher* and **Pad** the circles to a length of *20 mm*.

6) Put a **30° Draft Angle** on the surfaces of the cylinders.

7) Create a **15 mm Face To Face Fillet** between the two cones. (Unix users: Hold down the control key when selecting the 2 faces.)

2.5 - Variable Fillets:

Variable Fillets enable you to produce a fillet with a variable radius.

1) Create a **Variable Fillet** on the edge indicated that starts off at a **10 mm** radius and ends up at a **30 mm** radius using a **Linear** variation. Change the fillet radii by double clicking on the dimensions.

Filleted edge

10

30

2) Repeat on the other side using the **Cubic** Variation.

(Note: There are Drafted filleted pads and pockets available that do both commands simultaneously.)

Section 3: Creating Chamfers

1) Create a **3 mm** x **45°** Chamfer on the right face of the projection shown.

Chapter 3:
PART DESIGN

Tutorial 3.6: Dress-Up Features: Shell, Thickness & Threads

Featured Topics & Commands

Prerequisite Knowledge & Commands

- The *Sketcher* workbench and associated commands
- Editing a preexisting sketch
- Pads, Pockets, Shafts and Holes
- Fillets and Drafts
- Rotating the part using the mouse or compass

Shells

The *Shell* ![icon] command is part of the *Dress-Up Features* toolbar. Shelling consists of emptying a part out while keeping a given thickness on its sides. One or more faces may be selected to be open and the remaining sides can have different thicknesses.

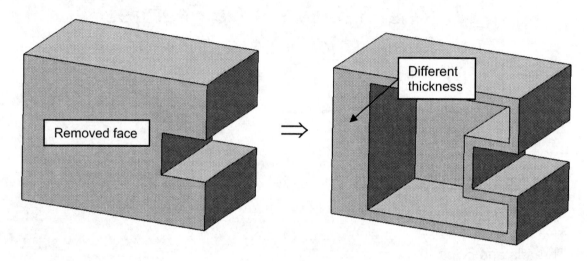

Thickness

The *Thickness* ![icon] command is part of the *Dress-Up Features* toolbar. This command allows you to add thickness to a wall.

Threads and Taps

The *Thread* ![icon] command is part of the *Dress-Up Features* toolbar. The *Thread* command allows you to create threads in accordance to drawing standards. User specified threads may also be implemented. The threads can not be viewed in the 3D space. However, they will appear on the orthographic projection of the part that is created in the *Drafting* workbench.

In the *Part Design* workbench, there is a command called *Tap –*

Thread Analysis ![icon] and is located in the *Analysis* toolbar. This command calls out all threads on the part. However, the call out that CATIA gives is not in accordance with ASME thread note standards.

Part Modeled

Three different views of the part modeled in this tutorial are shown below. The part is shown as a cut away view to show the inner features. The main solid is created using the *Shaft* command and is hollowed out using the *Shell* command. *Drafts, Fillets, Threads, Thickness* and *Holes* are added to finish the part. Notice that the threads cannot be seen in the 3D view, but are represented on the section view created in the *Drafting* workbench.

Thread symbol

Section view A-A
Scale: 1:1

Section 1: Applying a Shell.

1) Open a **New... Part** drawing, name it ***Threaded Support***, and enter the **Sketcher** on the **yz plane**.

2) Set your length units to be **inches**.

3) Draw and **Constrain** the following **Axis** and **Profile**.

4) **Exit** the *Sketcher* and **Shaft** the sketch ***360 deg***.

5) Pull out the *Dress-Up Features* toolbar .

6) **Shell** the part using a uniform thickness of *.625 in* and removing the bottom face of the part.

Remove this face.

7) **Edge Fillet** the edges indicated in the figure with a *.25 in* radius.

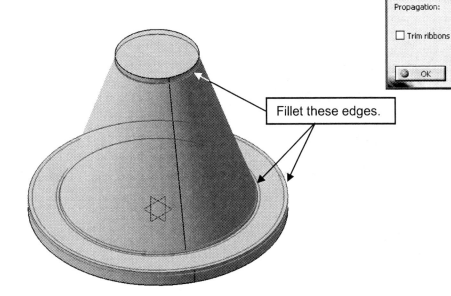

Fillet these edges.

8) **Edge Fillet** the edges indicated in the figure with a *.25 in* radius.

Fillet these edges.

Sketch face

9) Enter the **Sketcher** on the top face of the part.

10) Draw and **Constrain** the **Circle** shown. The circle's center coincides with the origin.

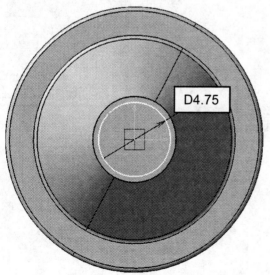

D4.75

11)**Exit** the *Sketcher* and **Pad** the sketch through the part *9 in*.

12)Change your view mode to **Wireframe** 🎞. This icon is located in the *View* toolbar located in the bottom toolbar area.

13)Put a **Draft Angle** 🗋 of *2 deg* on the padded cylinder that travels through your part using the top surface of the part as the *Neutral Element*. You may have to reverse the orange arrow so that the cylinder becomes smaller as it travels down.

Neutral Element (Top face)

Draft face

14) Change your view mode back to **Shading with Edges** .

15) Enter the **Sketcher** on the top face of the part.

Sketch face

16) Draw and **Constrain** the **Circle** shown. The circle's center coincides with the origin.

Face/Shaft.1/PartBody

D1.78

17) **Exit** the *Sketcher* and **Pocket** the sketch using the **Up to last** option.

Pocket Definition

First Limit

Type: Up to last

Limit: No selection

Offset: 0in

Profile/Surface

Selection: Sketch.3

☐ Thick

Reverse Side

☐ Mirrored extent

Reverse Direction

More>>

OK   Cancel

D 1.875

18) **Edge Fillet** the edge indicated in the figure with a *.5 in* radius.

Filleted edge

Section 2: Adding Thickness

1) Add .5 in of thickness to the top face of the part. Select the **Thickness** icon. Activate the *Default thickness faces:* field and select the top face of the part. Enter a thickness of *.5 in* in the *Default thickness:* field.

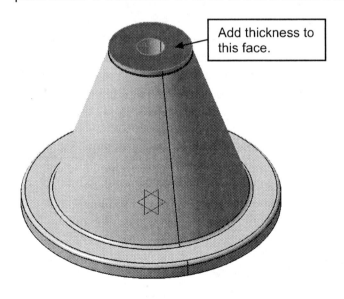

Add thickness to this face.

Section 3: Creating Threads

1) Apply a thread to the hole running down through the part. Select the **Thread** icon. Fill in the following fields; Lateral Face: Select the circumferential surface of the hole, Limit Face: Select the top surface of the part, Type: **No Standard**, Thread Diameter: *2 in*, Thread Depth: *9.5 in*, Pitch: *.222 in*, **Right-Threaded**. Notice that even though you can not see the threads, the specification tree contains the thread information.

2) Select the **Thread – Tap Analysis** icon in the *Analysis* toolbar. In the *Thread/Tap Analysis* window select the **More>>** button to see all the available options and then select **Apply**. Notice that a call out appears on the thread giving the thread information. Note that a more conventional thread call out for this particular thread would be **2 – 4.5 UNC – 9.5 LONG** where 4.5 is one over the pitch.

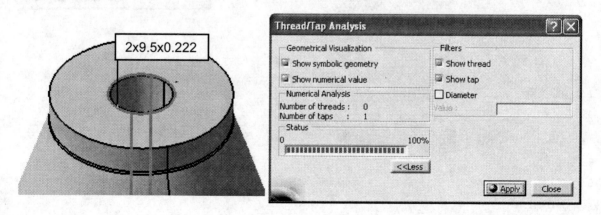

3) Create a Counterbored Hole in the bottom lip of the part with the following parameters; Counterbore diameter: **1 in**, Counterbore depth: **.125 in**, Hole diameter: **.625 in**. See hole location on the next page.

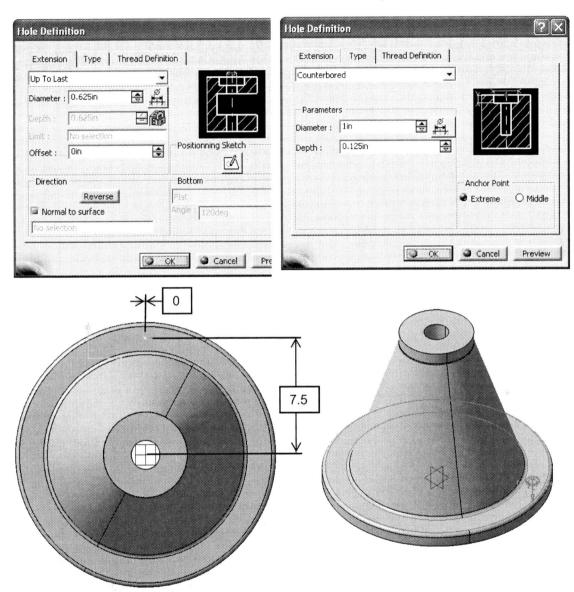

4) Create 3 more counter bored holes using a *Circular Pattern*. Select the

   Circular Pattern icon located in the *Transformation Features* toolbar. In the Circular Pattern Definition window, fill in the following fields; Object: Double click on the counterbored hole, Parameters: **Instance(s) & angular spacing**, Instance(s): *4*, Angular spacing: *90 deg*, Reference element: select the circumferential surface of the base, Object: select the counter bored hole on the part or in the specification tree.

Reference element

Chapter 3: PART DESIGN

Tutorial 3.7: Transformation Features

Featured Topics & Commands

Prerequisite Knowledge & Commands

- The *Sketcher* workbench and associated commands
- Editing a preexisting sketch
- Pads and Pockets
- Rotating the part using the mouse or compass

The Transformation Features toolbar

The *Transformation Features* toolbar contains commands that transform, move or create a pattern or array of a feature. Reading from left to right, the commands located in the *Transformation Features* toolbar are

- The Transformation toolbar: The *Transformation* toolbar contains commands that allow you to move, rotate and create a mirror image of a part.
- Mirror: The *Mirror* command creates a duplicate mirror image of a part.
- The Patterns toolbar: The *Pattern* toolbar allows you to create arrays of an existing feature.
- Scale: The *Scale* command changes the size of a part relative to a selected plane or point. Scaling can only occur in one direction at a time.

The Transformation toolbar

A transformation is the ability to move a body either by translating it along an axis, rotating it around an axis or moving it symmetrically about a plane. A body can not be duplicated during transformation. Reading from left to right, the commands located in the *Transformation* toolbar are

- Translation: *Translation* moves the part in a selected linear direction.
- Rotation: *Rotation* rotates a part around a selected line or axis.
- Symmetry: *Symmetry* replaces the part with its mirror image.

When transforming a part, CATIA will prompt you with a *Question* window to make sure it is okay to violate the constraints that were imposed in the *Sketcher*. If you are sure you want to continue with the transformation, click *Yes*.

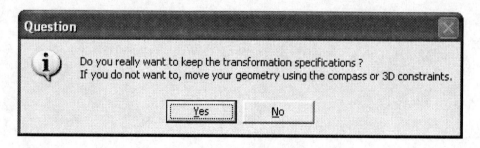

The Patterns toolbar

Patterns allow you to create arrays of an existing feature. You can create rectangular or polar arrays. The positions of the feature may also be manually defined. Patterned features are connected or associative. Therefore, if the original feature is changed all the patterned features will change. You can break the connection between the features by exploding the pattern. To explode the pattern, right click on the pattern feature in the specification tree and select *Pattern object - Explode...* Reading from left to right, the commands located in the *Patterns* toolbar are

- Rectangular Pattern: This command enables you to create rectangular arrays of a feature with a specified number of rows and columns.
- Circular Pattern: This command enables you to create polar arrays of a feature.
- User Pattern: This command allows you to create several instances of a feature in predefined locations. The locations are defined by the user in a sketch which usually consists of several points.

Part Modeled

The part modeled in this tutorial is a simplified representation of a bicycle chain ring. It contains some features that do not exist on a chain ring, but are created to illustrate a particular command. Ignoring the rectangular holes in the right arm, notice the parts symmetry and repeated features.

Section 1: Creating circular patterns.

1) Open a **New... Part** drawing, name it *Sprocket*, and enter the **Sketcher** on the **yz plane**.

2) Set your length units to be **inches**.

3) Draw and constrain the following sketch. Notice that the center of the arc is coincident with the origin.

4) **Exit** the Sketcher and **Pad** 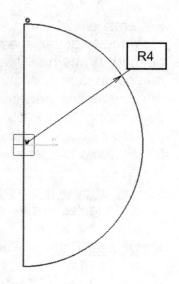 the sketch to a length of *.125 in*.

5) Enter the **Sketcher** on the front face of the Sprocket.

6) Draw and constrain the following **Construction Element** sketch.

7) **3D Project Elements** the face of the part on to the sketch plane as a **Construction Element**.

8) Using the construction lines as a guide, draw and constrain the sketch shown. The sides of the tooth are constructed using the **Profile** command and the tangent arc option. The top of the tooth is created using a **Three Point Arc**.

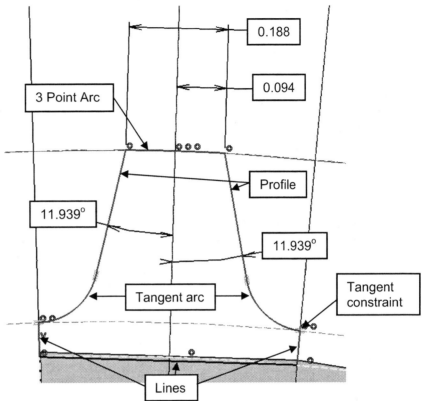

9) **Exit** the Sketcher and **Pad** the sketch to a length of *.125 in*. Reverse the direction if necessary.

10) Pull out the *Transformation Features* toolbar .

11) Create a circular pattern of the chain ring tooth. Select the **Circular Pattern** icon. In the Circular Pattern Definition window, fill in the following fields; Parameters: **Instance(s) & angular spacing**, Instance(s): **26**, Angular spacing: **6.923deg**, Reference element: Select the top of the tooth (You may have to zoom in to do this.), Object: select **Pad.2** (the pad used to create the tooth) in the specification tree. Reverse the direction if necessary.

Reference element for pattern.

Section 2: Mirroring a part.

1) Mirror your half chain ring to create a full one. Select the

   Mirror icon. Select the **zx plane** as your Mirroring element:. Read the prompt line for help.

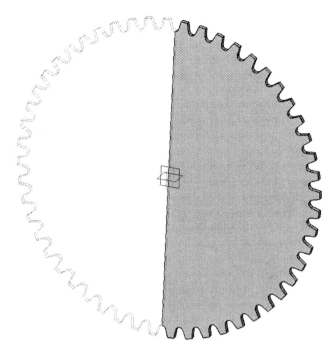

2) Enter the **Sketcher** on the **zx plane**.

3) Draw and **Constrain** the following **Rectangle**.

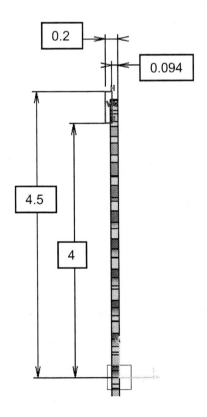

4) **Exit** the *Sketcher* and create a **Groove** using the sketch that travels **360 deg** using the **HDirection** as the Axis.

5) Enter the **Sketcher** on the front face of the part. Draw and constrain the following sketch. The length of the lines has to be long enough to go past the two circles, but the length beyond that doesn't matter.

6) Use the **Quick Trim** command to create a slot from the lines and circles previously draw.

7) Apply *0.09 in* radii **Corners** to the four sharp corners of the slot.

8) **Exit** the Sketcher and **Pocket** 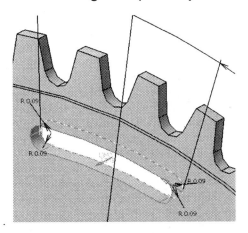 the sketch using the option **Up to last**.

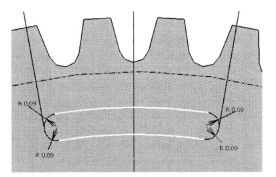

9) Create a **Circular Pattern** of the slot creating *12* instances spaced *30 deg* apart. Use the top of the tooth (as before) as the reference element.

10) Enter the **Sketcher** on the front face of the part. Draw and constrain the following sketch.

11) Use the **Quick Trim** command to create the two shapes shown from the lines and circles previously draw.

12) **Exit** the Sketcher and **Pocket** the sketch using the option **Up to last**.

13) Mirror Pocket.2. Select the **Mirror**

 icon and then select Pocket.2 in the specification tree. Select the **zx plane** as your Mirroring element:. Read the prompt line for help.

14) Enter the **Sketcher** on the front face of the part. Draw and constrain the following sketch.

15) **Exit** the Sketcher and **Pocket** the sketch using the option **Up to last**.

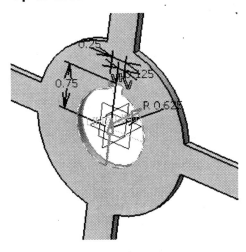

Section 3: Creating rectangular patterns.

1) Enter the **Sketcher** on the front face of the part. Draw and constrain the following sketch.

2) **Exit** the Sketcher and **Pocket** the sketch using the option **Up to last**.

3) Create a rectangular pattern of the rectangular slot that produces 2 rows 5 slots along the arm.

Select the **Rectangular Pattern** icon. In the *Rectangular Pattern Definition* window select the **First Direction** tab and fill in the following fields; Parameters; **Instance(s) & Spacing**, Instance(s): **5**, Spacing: **0.5 in**, Reference element: select the front face of the part, Object: **Pocket.4** (The pocket associated with the rectangular pocket). Select the **Second Direction** tab and fill in the following fields; Parameters; **Instance(s) & Spacing**, Instance(s): **2**, Spacing: **0.125 in**, Reference element: select the front face of the part, Object: **Pocket.4**.

4) Edit Sketch.7 (the sketch associated with the rectangular slot) and change the height of the rectangle from 0.063 in to **0.094 in**.

5) **Exit** the Sketcher. Notice that all the rectangular slots in the array update based on the original.

6) Right click on RectPattern.1 in the specification tree and select **RectPatter<u>n</u>.1 object – E<u>x</u>plode...** We have just broken the connection between the array elements. Notice that the RectPattern.1 feature is now broken up in to 10 individual pocket features.

7) Edit Sketch.7 and change the height of the rectangle back to **0.063 in**.

8) **Exit** the Sketcher. Notice that only the Sketch.7 updates. The other slots are unaffected.

Section 4: **Transformations.**

When performing the transformations in this section, click on *Yes* in the *Question* window when prompted to (If one appears).

1) Scale the thickness of the Sprocket by a factor of 2.

   Select the **Scale** icon and set the reference to be the **yz plane** and the ratio to be **2**.

2) Scale the part by *.5* using the **xy plane** as a reference.

3) **Rotate** the part by *45 deg* using the **X Axis** as the rotation axis.

4) Select the **Undo with history** icon. Undo all the transformations performed in this section.

NOTES:

Chapter 3: PART DESIGN

Tutorial 3.8: Boolean Operations

Featured Topics & Commands

Prerequisite Knowledge & Commands

- The *Sketcher* workbench and associated commands
- Editing a preexisting sketch
- Pads and Shafts
- Creating reference planes
- The *Transformation* toolbar
- Rotating the part using the mouse or compass

The Boolean Operations toolbar

Boolean operations allow you to create parts from individual solids. The commands located in the *Boolean Operations* toolbar, reading from left to right are

- <u>Add:</u> Adding a body to another one means uniting them and creating a single body.
- <u>Remove:</u> This command removes the intersecting volume of one body from another. The first body is then deleted.
- <u>Intersect:</u> This command removes everything except the volume that is created when two bodies intersect.

New part bodies

When you create a solid distant from your original pad, it is still considered to be the same part. If you move one, the other moves. In the specification tree, these two pads are under the same PartBody. In some instances, it is desirable to create a solid that can be moved and transformed independently of the other solids that already exist. In order to accomplish this, a new body has to be inserted. At the top pull down menu you would select *Insert – Body*. Now, any new solid that is created will be independent of the original PartBody.

When your part drawing includes several bodies, you can then associate these bodies in different ways (see the Assemble, Add, Trim, Intersect, Remove and Remove Lump commands) to obtain the final shape of the part.

Local Axis

The Axis System ⊥ command is used to create a coordinate system away from the three original coordinate planes. The origin and axis directions may be defined. This new coordinate system may also be rotated.

Part Modeled

The part modeled in this tutorial is shown below. It is created using the three main Boolean Operations (Add, Intersect, and Remove). New part bodies are inserted frequently to allow independent transformations.

Section 1: Inserting a new part body.

1) Open a **New... Part** drawing. If prompted, name it **Handle**.

2) Enter the **Sketcher** on the **zx plane**.

3) Set your length units to be millimeters (**Tools** – **Options...** – **Parameters and Measures**).

4) Draw and **Constrain** the **Circle** shown. The center of the circle is located at the origin.

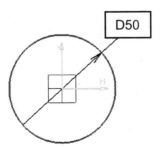

5) **Exit** the *Sketcher* and **Pad** the sketch ***30 mm*** in both directions.

6) At the top pull down menu, select **Insert** – **Body**.

7) Enter the **Sketcher** on the **yz plane**. Draw and **Constrain** the **Circle** shown. The center of the circle is located at the origin.

8) **Exit** the *Sketcher* and **Pad** the sketch ***30 mm*** in both directions. Your specification tree should look like the one shown.

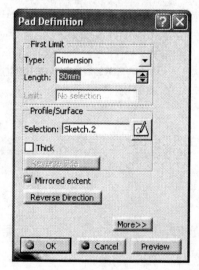

Section 2: Boolean operations.

1) Activate your Boolean operations toolbar if not already active. At the top pull down menu, select **View – Toolbars – Boolean Operations**.

2) Deselect all.

3) Select the **Intersect** icon and select the two cylindrical part bodies either in the tree or by selecting the bodies themselves. Select **OK** in the Intersect window.

4) At the top pull down menu, select **Insert – Body**.

5) Enter the **Sketcher** on the **zx plane**. Draw and **Constrain** the **Circle** shown. The center of the circle is located at the origin.

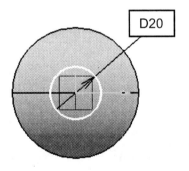

D20

6) **Exit** the *Sketcher* and **Pad** the sketch **70 mm** in both directions. Your specification tree should look like the one shown.

7) Select the **Translate** icon. Select **Yes** in the *Question* window. Translate Body.3 **10 mm** in the **Z Axis** direction.

8) Deselect all.

9) Select the **Add** icon and add **Body.3** to the original **PartBody**.

10) At the top pull down menu, select **Insert – Body**.

11) Enter the **Sketcher** on the **xy plane**. Draw and **Constrain** the **Oriented Rectangle** as shown.

12) **Exit** the *Sketcher* and **Pad** the sketch **30 mm** in both directions.

13) **Remove** Body.4 from the original PartBody.

14) At the top pull down menu, select **Insert – Body**.

15) Enter the **Sketcher** on the end of the shaft.

Sketch face

Face/Translate.1/Body.3

16) **3D Project Elements** the end of the shaft as a Construction Element.

17) Draw and **Constrain** the **Axis** and **Arc** shown.

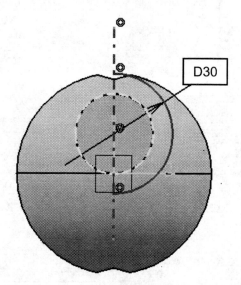

D30

18)**Exit** the *Sketcher* and **Shaft** the sketch *360°*.

19)**Translate** Body.5 (the sphere) *10 mm* in the *Y Axis* direction.

20)Insert a new Body.

21)Enter the **Sketcher** on the **yz plane**. Draw and **Constrain** the **Axis** and **Profile** shown.

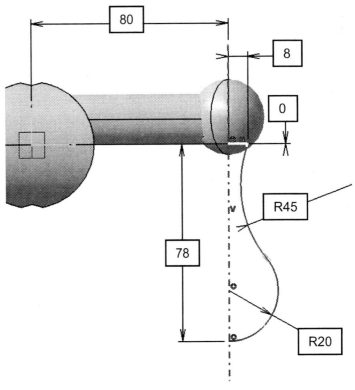

22) **Exit** the *Sketcher* and **Shaft** the sketch *360°*.

23) **Add** Body.5 to Body.6.

Section 3: Local axes.

1) Create a new **Plane** that is offset **80 mm** from the **zx plane**.

2) In the *Tools* toolbar, select the **Axis System** icon. In the *Axis System Definition* window, fill in the following fields; Y axis: select the new plane (Plane.1), Z axis: select the xy plane. (Note: The new axes will be perpendicular to the plane selected to define them.)

3) **Rotate** the handle grip (Body.5+Body.6) by **20°** using the newly created **X axis** as the rotation axis.

4) **Mirror** the handle using the **zx plane** as the mirroring element.

5) **Add** the handle (Body.5) to the main PartBody.

6) At completion, the specification tree should look similar to that shown on the right.

Chapter 3: PART DESIGN

Tutorial 3.9: Material Properties

Featured Topics & Commands

Prerequisite Knowledge & Commands

- The *Sketcher* workbench and associated commands
- Editing a preexisting sketch
- Pads
- Rotating the part using the mouse or compass

Applying and modifying material properties.

A specific material may be assigned to your part by using the *Apply Material* icon. There are many predefined materials available to choose from including metals, plastics, and woods. After a material has been assigned to the part, its properties may be modified to suit your particular needs. This is done by right clicking on the material in the specification tree and selecting *Properties*. Once the correct material properties have been assigned to your part, the part

properties may be obtained by selecting the *Measure Inertia* icon. This is located in the *Measure* toolbar. This command gives you information on the part's volume, area, mass, and moments of inertia. It is also used to modify how the part looks or is rendered, and the type of section line symbol that will be used in any section views that are created.

Part Modeled

The part modeled in this tutorial is simple. It is only used as a base for learning about applying and modifying materials and analyzing part properties.

Section 1: Applying Material.

1) Open a **New... Part** drawing. If prompted, name it **Block**.

2) Enter the **Sketcher** on the **yz plane**.

3) Before you start to draw make sure your length units are set to millimeters. (**Tools** – **Options...** – **Parameters and Measures**)

4) Draw and **Constrain** the **Rectangle** shown.

5) **Exit** the *Sketcher* and **Pad** the sketch to a length of **50 mm**.

6) At the top pull down menu, select **V̲iew – Render Style – C̲ustomize View**. In the *Custom View Modes* window, activate the **Materials** toggle.

7) Select **PartBody** in the specification tree and then select the **Apply Materials** icon. This is usually located in the bottom toolbar area. Select the **Metal** tab and then the **Steel** icon. Then **Apply Material** and select **OK**.

Section 2: Part properties.

1) Select **PartBody** in the specification tree and then the **Measure Inertia** icon. Usually located in the bottom toolbar area. Notice that the *Measure Inertia* window gives the part's volume, area, mass, center of gravity, and moments of inertia about the center of gravity. This information may be exported to a text file. For a simple part this is trivial information and can be calculated by hand. However, for more complex parts this information is not easily calculated by hand.

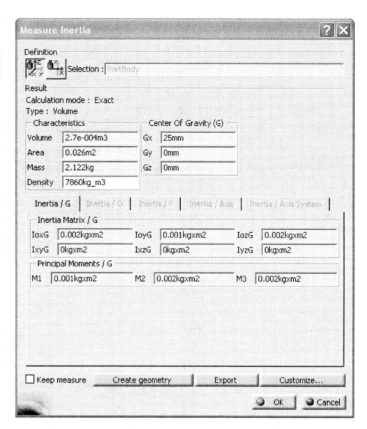

2) In the *Measure Inertia* window, select the **Customize...** icon. In the *Measure Inertia Customize* window, activate the **Inertia matirix / O** toggle and select **OK**. Now the moments of inertia about the origin are calculated.

(Note: Properties for a face may also be obtained by selecting only the face and then the *Measure Inertia* icon.)

Section 3: Modifying material properties.

1) In the specification tree, right click on **Steel** and select **Properties**.

2) In the *Properties* window, select the **Analysis** tab. Notice that you can change the material type between Isotropic and Orthotropic, and the structural properties to suit your specific steel type.

3) Select the **Drawing** tab. This is where you can set your section line (hatching) style if a section view is a planned part of your orthographic projection.

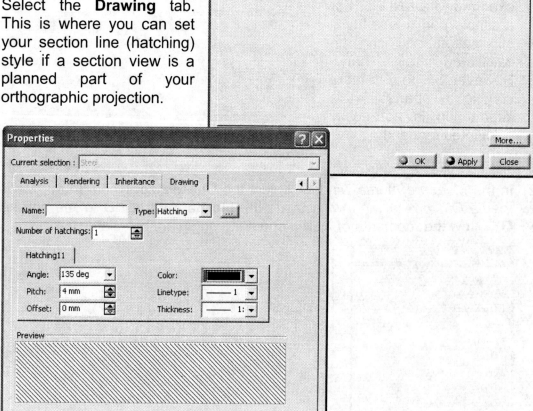

4) In your specification tree, right click on **Steel** and select **Delete**.

Section 4: Rendering

1) Select **PartBody** in the specification tree and then select the **Apply Materials** icon. This is usually located in the bottom toolbar area. Select the **Wood** tab and then the **Alpine Fir** icon. Then **Apply Material** and select **OK**.

2) In the specification tree, right click on **Alpine Fir** and select **Properties**.

3) In the *Properties* window, select the **Rendering** tab. Expand the **Change Mapping type and Preview** icon and select **Cubical Mapping**. Change the Material size: to *100 mm* and select **Apply**. Notice the difference. Select **OK**.

NOTES:

Chapter 3: Exercises

Exercise 3.1: This exercise can be performed after completing tutorial 3.1. Model the following parts.

a) Use at least one *Multiple Pad* when modeling this part.

b) Model this part using only one sketch and one *Multi-Pad* command.

Exercise 3.2: This exercise can be performed after completing tutorial 3.2. Model the following parts applying the appropriate constraints. Use the *Hole* command to create any holes, counterbores or threads.

a)

b)

Exercise 3.3: This exercise can be performed after completing tutorial 3.3. Model the following parts applying the appropriate constraints. Use the *Shaft* command to create the main solid.

a)

Ø100
⌴ Ø200
▽ 125

Ø400   Ø250   Ø250

100

225

250

NOTE: ALL FILLETS AND ROUND
R10 UNLESS OTHERWISE SPECIFIED

b)

2X Ø1.00
⌴ Ø1.75
▽ .38

Ø2.00   Ø3.50   Ø4.00

.12

.75

1.00

1.62

1.75

Exercise 3.4: This exercise can be performed after completing tutorial 3.4 and 3.5. Sketch the following parts applying the appropriate constraints. Apply *Edge Fillets* where necessary and at least on *Stiffener*. If you encounter problems applying an edge fillet, apply that one first before apply the others.

a)

NOTE: ALL FILLETS AND ROUNDS
RO.125 UNLESS OTHERWISE SPECIFIED

b)

R 45
R 64
83
∅ 70
4X ∅ 19
147
102
38
19
∅ 70
152
32
200

∅ 38
117
51
101
25

178
114
∅ 38
101
213

NOTE: ALL FILLETS AND ROUNDS R3
UNLESS OTHERWISE SPECIFIED

Exercise 3.5: This exercise can be performed after completing tutorial 3.7. Model the following parts applying the appropriate constraints. Use the appropriate transformations in the *Part Design* workbench.

a)

R 5

4

(5)

1

22 x 9.5 ELLIPSE
(AT RIM)

24 x 11 ELLIPSE
(AT HUB)

Detail D
2XScale

6X3 KEYWAY

⌀ 25

64

32

D

84

⌀ 178   ⌀ 50

2X 2 × 45°

NOTE: ALL FILLET AND ROUNDS R3
UNLESS OTHERWISE SPECIFIED

b)

NOTES:

Chapter 4: WIREFRAME AND SURFACE DESIGN FUNDAMENTALS

Introduction

Chapter 4 focuses on CATIA's *Wireframe and Surface Design* workbench. The reader will learn how to create basic wireframe surfaces and use them to create solids.

Tutorials Contained in Chapter 4

- Tutorial 4.1: Wireframe toolbar
- Tutorial 4.2: Surfaces toolbar
- Tutorial 4.3: Operations toolbar

NOTES:

Chapter 4:
WIREFRAME AND SURFACE DESIGN FUNDAMENTALS

Tutorial 4.1: Wireframe toolbar

Featured Topics & Commands

Prerequisite Knowledge & Commands

- Entering workbenches
- The *Sketcher* workbench and associated commands
- The *Part Design* workbench and associated commands

The Wireframe and Surface Design Workbench

The *Wireframe and Surface Design* workbench is used to create complex wire frame profiles and surfaces. In the *Part Design* workbench we worked under a *PartBody* which contained features used to create a solid. In the *Wireframe and Surface Design* workbench an *Open body* will contain features used to create surface elements. When creating reference elements (planes, lines, points) in *Part Design*, an *Open body* is automatically created to contain these elements. All the elements drawn in the *Wireframe and Surface Design* workbench will be under an *Open body*. Nothing will be solid. At any time you may insert an *Open body* by selecting *Insert – Open Body*.

The general process for creating a wireframe object is as follows: enter the *Wireframe and Surface Design* workbench, create wireframe geometry, create surfaces from existing geometry, modify and transform the surfaces (trim, join), enter the *Part Design* workbench, create a part body (solid), and modify the part using *Part Design* commands.

The *Wireframe and Surface Design* workbench contains the following standard workbench specific toolbars.

- <u>Wireframe toolbar:</u> This toolbar contains commands used to create reference elements such as points, lines, and curves that will be used to create surfaces.

- <u>Surfaces toolbar:</u> Once the reference geometry is constructed, the commands located in the *Surfaces* toolbar are used to create the surfaces.

- <u>Operations toolbar:</u> The commands located in the *Operations* toolbar are used to modify and transform surfaces and reference elements.

- <u>Replications toolbar:</u> The command located in the *Replications* toolbar are used to duplicate features such as points and planes.

The Wireframe toolbar

The commands located in the *Wireframe* toolbar are used to create reference elements such as points, lines, and curves that will be used to build surfaces. The commands and sub-toolbars located in the *Wireframe* toolbar, reading left to right, are

- The Points toolbar: Contains command to create reference points relative to the origin, another point, a curve or a surface.
- The LineAxisPolyline toolbar: Used to create lines, axes, and polylines.
- Plane: Allows you to create reference planes similar to that in the *Part Design* workbench. You can use planes as reference elements or cutting elements.
- Projection: This command creates a projection of a point, curve, or surface onto a support element, usually a plane or surface.
- Intersection: This command creates a curve that follows the intersection of a surface with a specified element.
- The Circle-Corner toolbar: Contains commands that will create circles, corners (fillets), and connect curves.
- The Curves toolbar: Used to create splines and helixes.

The Points toolbar

A point can be defined by its coordinates from a reference point (origin or selected point), or with respect to an element. You can edit any point by double clicking on it or its identifier in the specification tree. The commands located in the *Points* toolbar, reading left to right, are

- Point: The *Point* command allows you to create a point by entering coordinates relative to the origin or another point. It also allows you to create a point on a curve or surface relative to a point (end point, corner, reference point) on the curve or surface.
- Points Creation Repetition: The *Points Creation Repetition* command enables you to create several instances of a point.

The LineAxisPolyline toolbar

Lines can be used as guides, axes, or directions. They can also be joined to other elements. Lines can be created from points or vertices. They can be created on curves or surfaces. Lines have a direction associated with them depending on how you choose your start and end point. You can edit any line by double clicking on it or its identifier in the specification tree. The commands located in the *LineAxisPolyline* toolbar, reading left to right, are

- Line: The *Line* command allows you to create lines in various ways. A few examples are; between two points, from a point extending in a specified direction, and normal to a plane.
- Axis: The *Axis* command creates a reference element that is usually used to indicate symmetry.
- Polyline: The *Polyline* command allows you to create multiple lines that are connected together.

The Circle-Corner toolbar

The *Circle-Corner* toolbar contains commands that are used to create curved elements. The commands, reading from left to right, are

- Circle: Used to create circles. Unlike drawing a circle in the *Sketcher* workbench, a support plane or surface must be specified. If the surface is not flat, the circle will be projected normally onto the surface.
- Corner: This command creates a fillet between two elements.
- Connect Curve: This command creates a curve that will connect two elements.

The Curves toolbar

Curves are used as guides, reference elements, or as limits to surfaces. The commands located in the *Curves* toolbar are used to create complex curves. The commands, reading from left to right, are

- Spline: A spline is a fitted curve that connects several points. Tangencies may be specified at its ends.
- Helix: A helix is a spiral that travels along a specified axis.

Part Modeled

The part modeled in this tutorial is shown on the right. Reference elements (points, lines, planes, splines, ect...) are used to define surfaces. Once a surface model is created in the *Wireframe and Surface* workbench, the part is taken to the *Part Design* workbench and solids are created.

Section 1: Points and lines

1) Open a new part drawing and enter the *Wireframe and Surface Design* workbench .

2) Set your length units to be millimeters (**Tools – Options... – Parameters and Measures**)

3) Insert a point at the origin. Select the **Point** icon. In the *Point Definition* window fill in the following fields: Point type: Coordinates, X = *0 mm*, Y = *0 mm*, Z = *0 mm*, Reference Point: **Default (Origin)**.

4) Create a line that starts at the origin and travels

100 mm in the z direction. Select the **Line** icon. In the *Line Definition* window fill in the following fields: Line type: **Point-Direction**, Point: **Point.1** (the point at the origin), Direction: **Z Axis** (the line direction is perpendicular to the direction plane), Support: **Default (None)**, Start: *0 mm*, End: *100 mm*, Length type = **Length**.

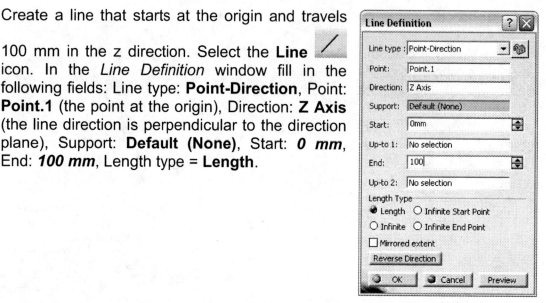

5) Create a point that is (-10, 30, -50) millimeters away from the top end of the line. Select the

Point icon. In the *Point Definition* window fill in the following fields: Point type: Coordinates, X = *-10 mm*, Y = *30 mm*, Z = *-50 mm*, Reference Point: **Line.1\Vertex** (The point at the top end of the line. This is not the end at the origin.).

6) Create points at (*0, 60, -70*), (*-10, -15, -50*), and (*0, -30, -70*) millimeters away from the same end of the line.

7) Create a line the starts at Point.3 and is

perpendicular to the line. Select the **Line** icon. In the *Line Definition* window fill in the following fields: Line type: **Angle/Normal to curve**, Curve: **Line.1**, Support: **yz plane**, Point: **Point.3**, Angle: *90 deg,* Start: *0 mm*, End: *60 mm*, Length type = **Length**.

8) Create a similar **Line** that starts at **Point.5**. Use the **Reverse Direction** icon if necessary.

Section 2: Splines

1) Create a **Spline** between **Line.1\Vertex**, **Point.2**, and **Point.3**. (Refer to figure on the previous page.) Create a similar **Spline** between **Line.1\Vertex**, **Point.4**, and **Point.5**.

2) Create four **Points** ■ that are located at (*50, -10, 0*), (*-50, -10, 0*), (*0, -10, 15*), and (*0, -10, -15*) millimeters relative to the origin.

3) Create a **Spline** 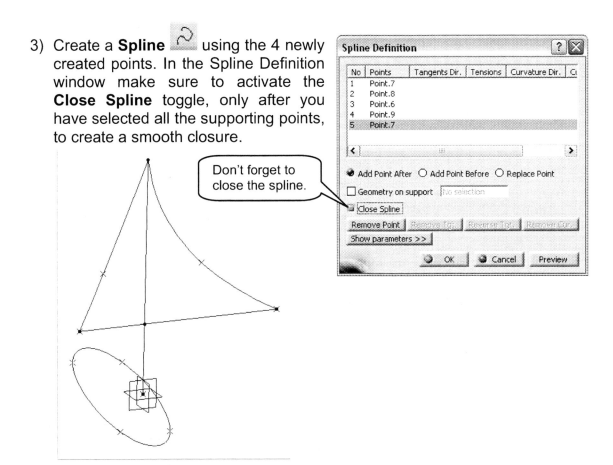 using the 4 newly created points. In the Spline Definition window make sure to activate the **Close Spline** toggle, only after you have selected all the supporting points, to create a smooth closure.

Don't forget to close the spline.

4) Draw a **Line** that starts at the origin and travels **100 mm** in the **y direction**.

5) Insert three points spaced 30 mm apart starting from the start of the newly created line.

Select the **Point** icon. In the *Point Definition* window fill in the following fields: Point type: **On curve**, Curve: **Line.4** (the newly created line), Distance to reference = **Distance on curve**, Length: *30 mm*, Reference Point:

Default (Extremity), and activate the **Repeat object after OK** toggle. After you select **OK** at *Points & Planes Repetition* widow will appear. In this window, fill in the following fields: Parameters: **Instances**, Instance(s): *2*, Second point: **Default (Extremity)**.

6) Create three planes each at the location of the three points just created and perpendicular to the line. Select the **Plane** icon. In the *Plane Definition* window fill in the following fields: Plane type: **Normal to curve**, Curve: **Line.4**, Point: **Point.10** (this is the point closest to the origin).

Create two more planes in a similar manner located at the remaining two points.

7) Project the ellipse located near the origin on to each of the newly created planes. Select the

Projection icon. In the *Projection Definition* window fill in the following fields: Projection type: **Normal**, Projected: **Spline.3**, Support: **Plane.1**. Follow a similar procedure for the remaining planes.

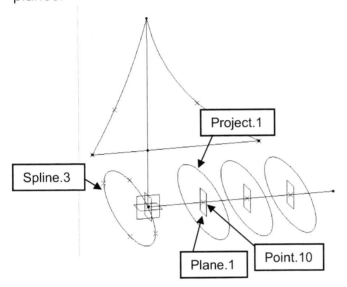

8) Scale the ellipse associated with Plane.1 by 80%. Select the **Scale** icon. This icon is located in the *Transformations* sub-toolbar which is located in the *Operations* toolbar. In the *Scaling Definition* window fill in the following fields: Element: **Projection.1**, Reference: **Point.10**, Ratio: *0.8*.

9) Using a similar process, scale the second projection by 60% and the third projection by 40%.

10) Insert points at the top of each of the three scaled

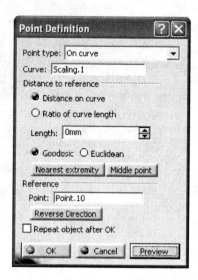

ellipses. Select the **Point** icon. In the *Point Definition* window fill in the following fields: Point type: **On curve**, Curve: **Scaling.1**, Distance to reference = **Distance on curve**, Length: *0 mm*, Reference Point: **Point.10**. Use a similar procedure to place a point on the top of the two remaining scaled ellipses. (Is the point at the bottom of the ellipse? Select the **Ratio of curve** length and enter *0.5 mm* in the ratio field.)

11) Create a Spline that connects the 5 points indicated in the figure.

Points used to create the spline

12) Draw a *2mm* radius **Circle** centered at the origin and supported by the **xy plane**. To make a complete circle click on the **whole circle** icon.

Section 3: Surfaces

1) Create surfaces for the sails. Select the **Fill**

    icon. This icon is located in the Surfaces toolbar. Select **Spline.1, Line.2,** and **Line.1**. Use a similar procedure to create a surface between **Spline.2, Line.3,** and **Line.1**.

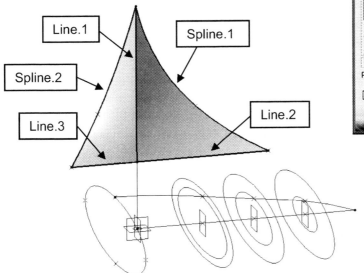

2) Activate the *Graphic Properties* toolbar. At the top pull down menu, select **View – Toolbars – Graphic Properties**. The *Graphic Properties* toolbar may be used to change the color of your surfaces.

3) Create a mast by sweeping the circle drawn at the origin along the vertical line. Select the

   Sweep 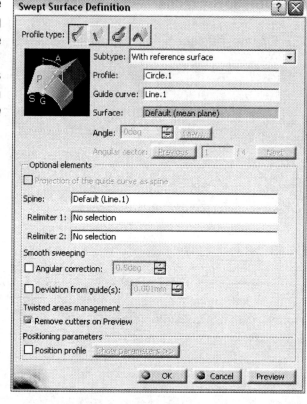 icon. This icon is located in the Surfaces toolbar. In the *Swept Surface Definition* window fill in the following fields: Profile: **Circle.1**, Guide curve: **Line.1**.

4) Use the **Fill** command to fill the top and bottom of the mast with a surface.

5) Use the ellipses and spline to create half of the surfboard. Select the

   Multisections Surface icon. Select the 4 ellipses starting with the original spline and selecting the other scaled ellipses in order of creation. In the *Multi-sections Surface Definition* window, select the **Spine** tab. Activate the Spine: field and select the spline that connects the 4 ellipses.

6) Use the **Fill** command to fill the small open end of the surfboard with a surface.

7) Use the **Plane** command to offset the **zx plane** *10 mm* in the negative y direction. The plane should be in the same location as the big open end of the surfboard.

8) Mirror the surfboard using the newly created plane as the mirroring surface.

Select the **Symmetry** icon. This icon is located in the *Transformations* toolbar. In the *Symmetry Definition* window activate the Element: field and select the surfboard (**Multi-sections Surface.1**). Activate the Reference: field and select the newly created plane (**Plane.4**).

9) Create a point on the top of the mirrored small end of the surfboard. Select the **Point** ■ icon. In the *Point Definition* window fill in the following fields: Point type: **On curve**, Curve: **Symmetry.1\Edge**, Distance to reference = **Distance on curve**, Length: *0 mm*, Reference Point: **Point.1**. (Is the point at the bottom of the ellipse? Select the **Ratio of curve** length and enter *0.5 mm* in the ratio field.)

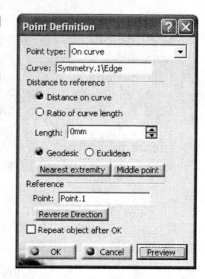

10) Create two **Points** ■ located at (*10, 20,-10*), and (*-10, 20, -10*) millimeters relative to the point at the top of the surfboard (the last point created).

11) Use the last three points created to define a reference plane. Select the **Plane** ⬭ icon. In the *Plane Definition* window select the Plane type: to be **Through three points** and then select the three points.

12) Use the newly created plane to create an intersection curve with the mirrored end of the surfboard. Select the **Intersection** icon. In the *Intersection Definition* window select the mirrored end of the surfboard (**Symmetry.1**) as the First Element:, and the newly created plane (**Plane.5**) as the Section Element:.

13)**Fill** 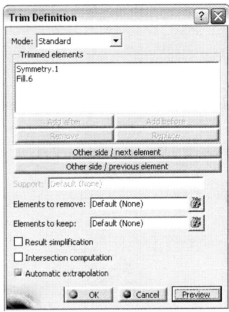 the intersection curve with a surface. (If the intersection curve does not fill, the most common reason is that the original spline used to create the initial ellipse was not closed. This usually produces a gap at the top of the intersection curve. Use a *Line* to connect the gap and try to *Fill* it again.)

14)Use the intersection surface to trim off the end of the surfboard. Select the **Trim** icon. This icon is located in the *Operations* toolbar. In the Trim Definition window select the mirrored end of the surfboard (**Symmetry.1**) as Element 1:, and the intersection surface (**Fill.6**) as Element 2:. (If the *Trim* command is trimming off the wrong side, select the *Other side* button.)

15)If the intersection curve lost its surface, **Fill** 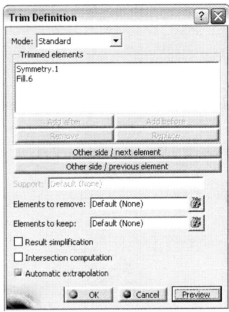 it again.

16)Create a point in the middle of the angled end of the surfboard (the intersection surface). Select the **Point** icon. In the *Point Definition* window fill in the following fields: Point type: **On surface**, Surface: select the angled end of the surfboard, Distance: **0 mm**, Reference Point: **Default (Middle)**.

17) Create a line that starts at the newly created point and travels out from surface 30 mm in a

normal direction. Select the **Line** icon. In the *Line Definition* window fill in the following fields: Line type: **Normal to surface**, Surface: select the angled end of the surfboard, Point: select the newly created point, Start: *0 mm*, End: *30 mm*, Length Type = **Length**. (You may have to use the Reverse Direction to get the line to travel away from the surfboard.)

18) Create a **Polyline** using the ends of the newly created line and the point indicated in the figure.

19) **Fill** the triangular curve with a surface.

20) **Join** the two halves and the filled ends of the surfboard.

Section 4: Creating solids

1) Enter the **Part Design** ⚙ workbench.

2) Activate the **Surface-Based Features (Extended)** toolbar if not already active.

3) Create a solid using the joined surface used to define the surfboard. Select the **Close Surface** icon. Select the surfboard and click **OK** in the *CloseSurface Definition* window. Notice that the surface of the surfboard now looks checkered. This is because both the solid and the surface now exist in the same space. Go to the specification tree and hide the surface used to define the surfboard (Join.1).

4) Use the **Close Surface** command to create a solid out of the swept mast. Then, hide the surface (Sweep.1).

5) We no longer need to see the reference elements (points, lines, planes, etc..). You may hide all of these if you like.

6) Now that you have solid, you may apply any *Part Design* workbench specific command to them.

Apply a *1.5 mm* radius **Edge Fillet** to the end of the surfboard that does not have the fin.

Chapter 4:
WIREFRAME AND SURFACE DESIGN FUNDAMENTALS

Tutorial 4.2: Surfaces toolbar

Featured Topics & Commands

Prerequisite Knowledge & Commands

- Entering workbenches
- The *Sketcher* workbench and associated commands
- The *Part Design* workbench and associated commands
- The *Wireframe* toolbar

The Surface toolbar

The commands located in the *Surfaces* toolbar are used to create surfaces that are defined by one or more reference elements. The commands, reading from left to right, are

- Extrude: An extruded surface is created from an open or closed curve. Its direction and limits are specified.
- Revolve: A revolved surface is created from an open or closed curve. The axis and angle of rotation are specified.
- Sphere: The *Sphere* command allows you to create a complete or incomplete sphere. Limits for the sphere may be specified in both the revolved direction (meridian) and the top to bottom direction (parallel).
- Cylinder: The *Cylinder* command allows you to create a complete or incomplete cylinder.
- Offset: An offset surface is created from an existing surface. An offset distance and direction are specified. The resulting surface is parallel to the existing one.
- Sweep: A swept surface is created from an open or closed curve that is swept along guide curves.
- Fill: The filled surface is created between joined curves. The shape of the resulting surface depends on the continuity between the support surfaces (if any) and the fill surface.
- Multisections Surface: A multisections surface or lofted surface is defined by several parameters including; tangency, closing point, coupling, and guide curve.
- Blend: The blend surface is created between two curves. The curves used must be single edge curves and not closed. The shape of the resulting surface depends on the continuity between the support surfaces (if any) and the blend surface.

Part Modeled

The part modeled in this tutorial is created with reference elements and surfaces. The surfaces are created in a variety of ways in order to gain experience with as many commands in the *Surfaces* toolbar as possible.

Section 1: Multisections surface (Loft)

1) Open a **New... Part** drawing, name it **Hammer** and enter the *Wireframe and Surface Design* workbench .

2) Set your length units to be inches (**Tools – Options... – Parameters and Measures**)

3) Create 5 **Points** that are located at (**0.5, 0, 0**), (**-0.5, 0, 0**), (**0, 0, 0.625**), (**0, 0, -0.625**), and (**0, 11, 0**) relative to the **origin**.

4) Create 4 **Points** that are located at (**0.3125, 0, 0**), (**-0.3125, 0, 0**), (**0, 0, 0.4375**), (**0, 0, -0.4375**) relative to the **Point.5** (the point located 11 inches from the origin).

5) Use a **Spline** to connect the first 4 points and last 4 points as shown in the figure. Start and end the spline at the top and activate the **Close spline** toggle.

Start and end spline at the top.

Point.5

6) Create 12 **Points** that are located at (**0, 2, 0.594**), (**0, 4, 0.531**), (**0, 6, 0.5**), (**0, 8, 0.406**), (**0, 9, 0.375**), (**0, 10, 0.344**), (**0.4375, 2, 0**), (**0.406, 4, 0**), (**0.375, 6, 0**), (**0.3125, 8, 0**), (**0.25, 9, 0**), (**0.219, 10, 0**) relative to the **origin**.

7) Use a **Spline** to connect the first 6 points and last 6 points as shown in the figure.

8) Use a multisections surface to loft together the two ellipses using the two long splines as guides. Select the **Multisections Surface** icon. In the Multi-sections Surface Definition window, activate the *Section* area and select the two ellipse (Spline.1 and Spline.2). Activate the *Guide* area and select the two long splines (Spline.3 and Spline.4). Make sure both closing points are at the top.

Section 2: Extruded surface

1) Create 4 **Points** ▪ that are located at (**0.25, 0.5, 0.375**), (**-0.25, 0.5, 0.375**), (**0.25, 0.5, -0.375**), (**-0.25, 0.5, -0.375**) relative to the **Point.5** (the point located 11 inches from the origin).

2) Draw 4 **Lines** ╱ connecting the last 4 points drawn as shown in the figure.

3) Create rounded corners or fillets at each vertex of the rectangle. Select the

Corner icon located under the *Circle* icon. Select two adjoining lines to be Element 1: and Element 2:. Activate both **Trim element** toggles and enter a Radius: of *0.19 in*. Repeat this procedure for the next two corners. On the last corner you need to activate the **Corner On Vertex** toggle.

4) **Extrude** the cornered rectangle by *1 in* along the **Y Axis**.

Section 3: Blended surface

5) Create a surface that connects Multi-sections surface.1 and Extrude.1. Select the

Blend icon. In the *Blend Definition* window, activate the *First curve* field and select the spline used to create the end of Multi-sections surface.1 (**Spline.2**). Activate the *Second curve* field and select the cornered rectangle. Note where the *First Point* and *Second Point* lie. They should be across from each other to prevent twist. If your *Second Point* is not across from your *First Point,* click on the **Closing Points** tab and **Replace** your *Second Point* with a point that is across from the *First Point.*

Need to move the Second Point to be across from the First Point.

The Second Point is now across from the First Point.

6) Notice that the interface between Blend.1 and Extrude.1 is discontinuous or sharp. Make that interface smooth. Get back into the *Blend Definition* window by double clicking on **Blend.1** in your specification tree. Activate the *Second support* field and select **Extrude.1**.

7) **Fill** the two ends of the handle.

8) **Join** all surfaces of the handle.

Section 4: Spheres and Cylinders

1) Insert a new Open Body (**Insert – Body**).

2) Insert a new **Plane** that is offset from the **zx plane** by **11.75 in**.

3) Insert a **Point** ▪ at (*0, 11.75, 0*) inches relative to the origin. Then insert a

Point ▪ at (*0, 0, -1*) inches relative to the above point.

4) Select the **Cylinder** ⬭ icon. In the *Cylinder Surface Definition* window fill in the following fields; Point: **Point.27** (most recently created point), Direction: **Z Axis**, Radius: *0.5 in*, Length 1: *0 in*, Length 2: *0.875 in*.

5) Insert a **Point** at (**0, 0, 1**) inches relative to the point in the middle of the offset plane (Point.26).

Point.26

6) Select the **Sphere** icon. In the *Sphere Surface Definition* window, fill in the following fields; Center: **Point.28** (most recently created point), Sphere radius: **0.438 in**, select the incomplete sphere icon, Parallel Start Angle: **0 deg**, Parallel End Angle: **90 deg**, Meridian Start Angle: **0 deg**, Meridian End Angle: **360 deg**.

7) **Fill** the bottom of the cylinder. The end that is furthest away from the handle.

Section 5: Revolved surface

1) Insert 2 **Points** 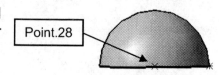 at (**-0.5, 0, 0**) and (**-0.44, 0, 0.25**) inches relative to the center point of the cylinder (Point.27).

2) Insert 2 **Points** at (**-0.438, 0, 0**) and (**-0.25, 0, -0.25**) inches relative to the center point of the sphere (Point.28).

3) Insert a **Point** at (**-0.5, 0, 0**) inches relative to the point located in the middle of the offset plane (Point.26).

4) Create a **Spline** connecting the last 5 points created as shown in the figure.

5) Create a **Line** that connects the center of the cylinder with the center of the sphere.

6) **Revolve** the spline *360 deg* about the line.

7) Create an **Intersection** curve between the revolved surface and the handle. When you select the **OK** button a *Multi-Result Management* window will appear. This means that two intersections were created and we need to tell CATIA which one to keep. After you select the **Yes** button, a *Near Definition* window will appear. CATIA wants you to choose an element that is near the intersection curve that you want to keep. Select an edge of the handle end.

Select an edge near the intersection to keep.

8) Hide the handle. Right click on **PartBody** and select **Hide/Show**.

9) **Extrude** the intersection curve (**Near.1**) *1 inch* in both directions along the **Y Axis**.

10) **Trim Revolute.1** and **Extrude.2** to create a hole going through the hammer head. Select the **Other side / previous elements** buttons until the desired result is obtained.

11) **Join** all the surfaces of the hammer head.

12) Show the hammer handle and hide all reference elements.

Section 6: Creating solids

1) Enter the **Part Design** workbench.

2) Right click on **PartBody** and select **Define in Work Object**.

3) Use the **Close Surface** command to create a solid out of the handle (Join.1).

4) Hide the surfaces of the handle (Join.1).

5) Apply a *0.06 in x 45 deg* **Chamfer** to the end of the hammer handle. The end away from the hammer head.

6) Right click on **Body.2** and select **Define in Work Object**.

7) Use the **Close Surface** command to create solids out of the hammer head (Join.2).

8) Hide the surfaces of the hammer head (Join.2).

9) Apply a **_0.06 in_** radius **Fillet** to the striking surface of the hammer head and the edge of the ball, as shown in the figure.

NOTES:

Chapter 4:
WIREFRAME AND SURFACE DESIGN FUNDAMENTALS

Tutorial 4.3: Operations toolbar

Featured Topics & Commands

Prerequisite Knowledge & Commands

- Entering workbenches
- The *Sketcher* workbench and associated commands
- The *Part Design* workbench and associated commands
- The *Wireframe* and *Surfaces* toolbar

The Operations toolbar

The *Operations* toolbar contains commands the will help you transform and modify surfaces and curves and make it easier to produce the desired shape. The commands and sub-toolbars located in the *Operations* toolbar, reading left to right, are

- Join-Healing toolbar: Contains commands that will join, fill gaps, restore, and disassemble adjoining elements.
- Split-Trim toolbar: Contains commands that will cut one side of intersecting elements.
- Extracts toolbar: Contains commands that create curves and surfaces from surface boundaries and edges.
- Transformations toolbar: Contains commands that will transform the size, position, orientation, etc... of an element.
- Extrapolate: The *Extrapolate* command extends one element to the boundary of another.

The Join-Healing toolbar

The commands contained in the *Join-Healing* toolbar, reading left to right, are

- Join: The *Join* command allows you to join one or more adjacent elements into a single element. Multiple curves or surfaces may be joined. You can define a *Merge distance* and an *Angle tolerance*. The *Merge distance* is the maximum gap distance under which the two elements can be joined. The *Angle tolerance* specifies the maximum interface angle under which the two elements can be joined.
- Healing: The *Healing* command fills in gaps between two elements. A *Merge distance* or maximum gap that will be filled, and a *Distance objective* or minimum gap that will be filled, may be specified.
- Untrim Surface or Curve: The *Untrim Surface or Curve* command allows you to rebuild an element that has been split one or more times.

- Disassemble: The *Disassemble* command breaks a multi-celled surface or curve into its individual cells.

The Split-Trim toolbar

The commands contained in the *Split-Trim* toolbar, reading left to right, are

- Split: The *Split* command cuts or splits one element using a cutting element. One side of the element is cut.
- Trim: The *Trim* command cuts or trims two intersecting elements. One side of each element is cut.

The Extracts toolbar

The commands contained in the *Extracts* toolbar, reading left to right, are

- Boundary: The *Boundary* command creates a curve from a surface boundary. Limits may or may not be set.
- Extract: The *Extract* command creates a curve from a surface edge. It can also be used to extract a surface. Limits may not be set.

The Transformations toolbar

The commands contained in the *Transformations* toolbar, reading left to right, are

- Translate: The *Translate* command makes a duplicate element that is a specified distance away from the original. Multiple duplicates may be created using the repeat option.
- Rotate: The *Rotate* command makes a duplicate element that is rotated about an axis by a specified angle. Multiple duplicates may be created using the repeat option.
- Symmetry: The *Symmetry* command mirrors an element about a reference element.
- Scaling: The *Scaling* command makes a duplicate element that is of a different size from the original. Multiple duplicates may be created using the repeat option.
- Affinity: The *Affinity* command creates a scaled version of an element. However, the scale factors in each direction may be different.
- Axis to Axis: The *Axis to Axis* command takes an element that is supported by one axis and creates a duplicate that is supported by another axis.

The Grid toolbar

The *Grid* toolbar is located in the *Tools* toolbar. The commands located in the Grid toolbar, reading left to right, are

- Work on Support: This command allows you to work on a sketch plane.
- Snap to Point: This command allows you to snap to the grid points.
- Working Supports Activity: This command exits the working support mode.

Part Modeled

The part modeled in this tutorial is a combination of shapes used to illustrate the commands found in the *Operations* toolbar. The reference elements used to constructed the base where drawn using the *Work on Support* command. Several commands found in the *Operations* toolbar and its sub-toolbars are used to modify and add to the base.

Section 1: Using Supports

1) Open a **New... Part** drawing, name it **Sheet Metal** and enter the *Wireframe and Surface Design* workbench .

2) Set your length units to be millimeters (**Tools – Options... – Parameters and Measures**)

3) Select the **Work on Support** 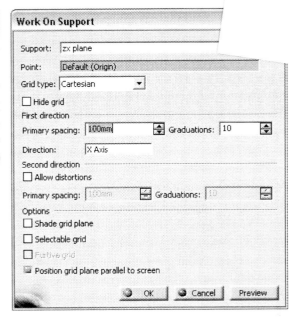 icon located in the *Tools* toolbar which is usually in the bottom toolbar area. In the first *Work on Support* window select the **zx plane** as the Support:. In the second *Work on Support* window set the Primary spacing to be *100 mm* and the Graduations: to be **10**.

4) Activate the **Snap to point** icon.

5) Draw the following profile using the **Polyline** command.

6) Select the **Works Support Activity** icon to deactivate the support. Then, select the **Isometric View** icon.

7) **Extrude** the polyline **75 mm** in the positive and negative **y** directions.

8) Use the **Work on Support** command to sketch on the **xy plane** with the same grid spacing as before.

9) Hide Extrude.1.

10) Draw the following 2 **Splines** using the points indicated in the figure.

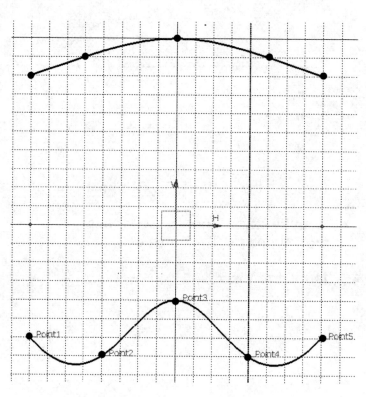

11) Unhide Extrude.1.

12) Select the **Works Support Activity** 🔲 icon to deactivate the support. Then, select the **Isometric View** 🔲 icon.

13) **Extrude** 🔲 each spline **20 mm** in the both the negative and positive **z** directions.

Section 2: Split, Trim, and Extrapolate

1) Select the **Split** 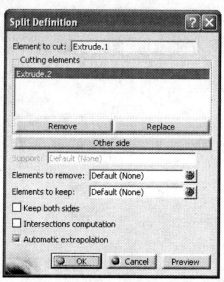 icon. In the *Split Definition* window, activate the Element to cut: field and select **Extrude.1**. Activate the Cutting elements area and select the extrude created with the wavy spline. Use the **Other side button** if necessary to get the desired results. Notice that the cutting element does not change.

2) Delete Split.1 and select the **Trim** 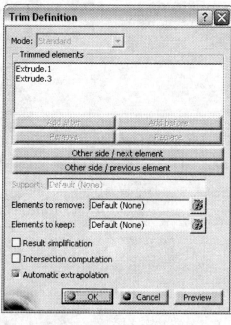 icon. Use Extrude.1 as Element 1: and the wave spline extrude as Element 2:. Use the **Other side of elements** buttons if necessary. Notice that both elements get cut.

3) Create a boundary to use in the *Extrapolate* command. Select the **Boundary**

icon. In the *Boundary Definition* window, fill in the following fields; Propagation type: **Point continuity**, Surface edge, Select the original **extrude** now called **Trim.1**, *Limit 1:* select **point 1** as shown on the figure, Limit 2: select **point 2**. We want the boundary to cover the original polyline shape, if it does not, click on the arrow attached to point 1.

4) Select the **Extrapolate** icon. In the *Extrapolate Definition* window, fill in the following fields; Boundary: **Boundary.1**, Extrapolate: **Trim.1**, Type: **Up to element**, Up to: select the extrude created from the curved spline.

5) Use the **Split** command to achieve the results shown in the figure.

Section 3: Extracting

1) Select the **Extract** icon and select the face indicated in the figure.

Select this face.

Extract Definition

Propagation type: No propagation

Element(s) to extract Extrapol.1\Face.1

Support : No selection

☐ Complementary mode

☐ Federation

Show parameters >>

OK   Cancel   Preview

2) Select the **Extract** icon repeatedly and select the edges indicated in the figure.

Select these edges one at a time.

3) Use the **Boundary** command to extract the remaining edge. You need to select the surface and then use limits to modify the boundary edge.

Use *Boundary* to extract this edge.

4) **Join** all the edges created in the previous 2 steps.

5) **Extrude** **Extract.1** and **Join.1** individually *40 mm* in the **z** direction.

6) Disassemble the closed extrude so that one of its surfaces can be extruded. Select the closed extrude (Extrude.4 or Extrude.5 depending on the order in which you extruded) then select the **Disassemble** icon. A Disassemble window will appear showing how many elements the extrude will be broken up into. Select **OK**.

7) **Extrude** the back surface of the disassemble extrude **75 mm** in the **y** direction as shown in the figure. You may have to select the surface in the specification tree.

8) Use the Affinity command to create a scaled duplicate of the open extrude. Select the **Affinity** icon. In the *Affinity Definition* window, fill in the following fields; Element: select the open extrude (Extrude.4 or Extrude.5), Origin: select the point indicated in the figure, X: **0.5**, Y: **-1.5**, Z: **0.5**.

NOTES:

Chapter 4: Exercises

Exercise 4.1: This exercise can be performed after completing the tutorials presented in chapter 4. Model the following parts.

    a) A pair of pliers
    b) A screwdriver
    c) A popup tent
    d) An airplane body
    e) A car body

NOTES:

Chapter 5: ASSEMBLY DESIGN FUNDAMENTALS

Introduction

Chapter 5 focuses on CATIA's *Assembly Design* workbench. The reader will learn how to create and manage an assembly or product, and apply assembly constraints.

Tutorials Contained in Chapter 5

- Tutorial 5.1: Assembly Constraints
- Tutorial 5.2: Advanced Commands

NOTES:

Chapter 5:
ASSEMBLY DESIGN FUNDAMENTALS

Tutorial 5.1: Assembly

Featured Topics & Commands

Prerequisite Knowledge & Commands

- Entering workbenches
- The *Sketcher* workbench and associated commands
- The *Part Design* workbench and associated commands
- Moving and rotating parts with the compass

Assembly Design Workbench

The *Assembly Design* workbench allows you to design in the context of an assembly. You can add new or existing parts, or sub-assemblies to the root assembly. Parts can be positioned and constrained within the assembly. You can also analyze interferences between parts.

Product Structure

The product structure is best visualized by looking at the specification tree. Let's take, for example, the assembly of a machine vise. At the root is the product or assembly (Machine Vise). Under the assembly are parts (Base, Sliding Jaw, Jaw Plate, etc...), constraints that are used to place parts in their functional positions, and a subassembly (Handle Assembly). Notice that there are multiple copies of a single part. For example, there are 4 instances of Part11. Part11 was not drawn four different times; it was merely copied and pasted using the familiar CTRL+C and CTRL+V shortcuts. The root assembly and parts can be named by right clicking on the name and accessing the *Properties* window.

The Product Structure Tools toolbar

This toolbar allows you to insert, replace, and reorder components in your product. When you are inserting an existing part or subassembly, their corresponding files are not copied into the assembly. They are just referenced by the assembly. The commands located in the *Product Structure Tools* toolbar, reading left to right, are

- Component: This command inserts a new component that exists only in the root assembly file (CATProduct) and not as its own file.
- Product: This command allows you to insert a new product or subassembly (CATProduct) into the root assembly.
- Part: This command allows you to insert a new part (CATPart) that will be designed on the fly while you are designing the assembly.
- Existing Component: This command is used when you want to add an existing part or assembly as a component.
- Existing Component With Positioning: This command is used when you want to add an existing part or assembly as a component with a position.
- Replace Component: This command is used when you want to replace a component.
- Graph Tree Reordering: This command allows you to reorder your components in the specification tree.
- Generate Numbering: This command allows you to associate numbers or letters to each part of the assembly. These numbers or letters will be used when creating and ballooning the assembly drawing in the *Drafting* workbench.
- Selective Load: This command tells which components in the assembly will be loaded (opened or active) or not, and components that will be hidden or not.
- Manage Representations: This command allows you to manage the representations (name, association, etc...) of the components in the assembly.
- Multi Instantiation toolbar: This toolbar contains commands that allow you to create instances or duplicates of a component.

The Catalogue toolbar

The *Catalogue Browser* command lets you browse through existing already drawn catalogue of standard parts such as bolts, screws and nuts. Therefore, there is no need to model these components if they can be retrieved from the catalogue.

The Constraints toolbar

Assembly constraints are the constraints used to place each part into the assembly in its functional position. The general process for constraining the parts in an assembly is; fix one component in space, use the compass to move the other parts into approximate position, position the parts precisely using the correct constraint(s), and then update the assembly to move the parts into position. If a part(s) has been moved out of position, the update command may be used to move the part(s) back into position. Constraint symbols may be hidden by right clicking on the *Constraints* inference in the specification tree and selecting *Hide/Show*. The commands located in the Constraints toolbar, reading left to right, are

- <u>Coincidence Constraint:</u> The *Coincidence Constraint* creates alignments between axes, planes, or points.
- <u>Contact Constraint:</u> The *Contact Constraint* creates contact between two planes or faces.
- <u>Offset Constraint:</u> The *Offset Constraint* defines an offset distance between two elements.
- <u>Angle Constraint:</u> The *Angle Constraint* defines an angle between two elements.
- <u>Fix Component:</u> This command fixes a part in space such that it is always located in the same spot. The other parts are constrained relative to the fixed part.
- <u>Fix Together:</u> This command fixes two parts together.
- <u>Quick Constraint:</u> This command automatically creates what CATIA believes is the correct constraint between two elements.
- <u>Flexible/Rigid Sub-Assembly:</u> This command allows you to change a sub-assembly between flexible and rigid. A flexible sub-assembly has part or components that can be moved disregarding the fact that it is not the active component. The positions of the parts can be different than those stored in the reference CATProduct file.
- <u>Change Constraint:</u> This command allows you to change the type of constraint that exists between two elements.
- <u>Reuse Pattern:</u> This command allows you to reuse an existing pattern.

Save Management

Since assemblies contain several file links, the *Save Management...* command should be used to save it. A *Save Management* window will appear showing all the CATProducts and CATParts that have been or need to be saved. You can click on each individually and save the file using the *Save* or *Save As* commands. Also, the *Save All* command will save all documents that have been modified since the last save.

Assembly/Parts Modeled

The assembly and parts modeled in this tutorial are shown in the figures below. The main objective of this tutorial is to familiarize you with the product structure and assembly constraints. However, assemblies are made up of parts that must fit together and fasten. Therefore, the parts modeled in this tutorial have toleranced dimensions and threads.

PART #	PART NAME	REQ'D	MAT
1	TOOL POST BODY	1	CI
2	WEDGE	1	SAE 1045
3	BOLT	1	SAE 1040
4	TOOL POST SCREW	2	SAE 1040
5	ADJUSTING SCREW	1	SAE 1040
6	HEX NUT	1	–
7	SET SCREW	2	SAE 1040
8	WASHER	1	SAE 1040
9	WASHER	1	SAE 1040
10	ADJUSTING NUT	1	SAE 1040

Section 1: Setting up your product

1) Enter the **Assembly Design** workbench. Notice that at the top of your specification tree there is a Product.

2) Set your units to be **Inches** (**T**ools – **O**ptions...).

3) Right click on **Product1** and select **Pr**o**perties**. In the *Properties* window, click on the **Product** tab. In the Part Number field, rename your product **Tool Post**.

4) Save your draw as **Tool Post.CATProduct**.

Section 2: Part 1 (Tool Post Body)

NOTE: ALL FILLETS AND ROUNDS R.06
UNLESS OTHERWISE SPECIFIED

1) Select the **Part** icon (If **Part1** does not appear in the specification tree, then click on **Tool Post**). This will add a new part to your assembly or product. Notice that *Tool Post* in the specification tree is highlighted in blue. This means that it is the active component.

2) Right click on **Part1 (Part1.1)** in the specification tree and select **Properties**. In the *Properties* window, click on the **Product** tab. In the Instance name field, rename your part **Tool Post Body**.

3) At the top pull down menu, select **File – Save Management....** A *Save Management* window will appear indicating that *Tool Post.CATProduct* has been saved before, but modified since, and that *Part1.CATPart* is new. Click on **Tool Post.CATProduct** and select the **Save** button. Then, click on **Part1.CATPart** and select the **Save As...** button. Name your file **Part1 – Tool Post Body.CATPart**.

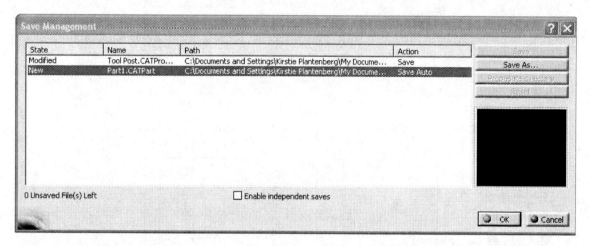

4) Expand *Part1 (Tool Post Body)* in the specification tree until you can see its *PartBody*. Double click on **PartBody**. This will automatically enter you into the *Part Design* workbench. You are now ready to draw Part1. Notice that now Part1 is highlighted in blue. If CATIA does not enter the *Part Design* workbench (it may enter the *Wireframe* workbench if this was the last one you were in), you will have to enter it manually.

5) Draw the right side profile of Part1 on the **zx plane** as shown and *Pad* it to the appropriate length.

6) Fillet the appropriate edges.

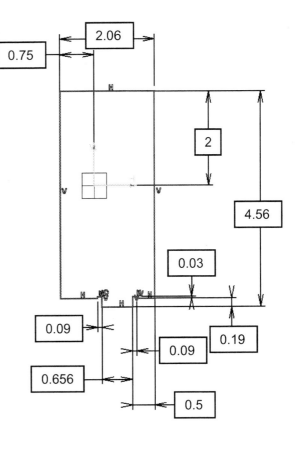

7) Draw and *Pocket* the slot in the front of Part1. Hint: Projecting the part edges will help.

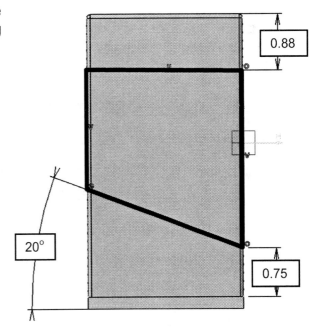

8) Create the *Hole* that travels from top to bottom. Notice that this hole is toleranced. Position the sketch before filling in the information. The command tends to lock up if you position your sketch at the end. To access the *Limit of Size Definition* window click on the tolerance icon. Note, the tolerance annotation shown on the CATIA part will be in metric (ϕ12.7 +0.1 -0.05)

9) Create the two threaded *Holes* on the top. The thread note is 7/16 – 14 UNC – 3B. The Thread Diameter: (major diameter) is 7/16, the Hole Diameter: (tap drill) is 0.368 as obtained from a thread table, and the Pitch: is 1/14.

10)Create the two threaded *Holes* in the front.

11)Create the Counter bored *Hole* in the side.

12) After completion, use **Sa̲ve Management...** to save your product and part. Your part and specification tree should look like this.

Section 3: Part2 (Wedge)

.75

3.50

1.62

.280

.190
.188

70°

1) Enter the *Assembly Design* workbench by double clicking on **Tool Post**.

2) Select the **Part** icon and then click on **Tool Post** in your specification tree. You can select **N̲o** in the *New Part: Origin Point* window.

3) Hide *Part1 (Tool Post Body)* by right clicking on it and selecting **Hide/Show**.

4) Rename your new part **Wedge**.

5) Expand *Part2 (Wedge)* and double click on its **PartBody**. You are in the *Part Design* workbench and ready to draw the Part2.

6) Model the Wedge. Notice that one of its dimensions is toleranced (assume the basic dimension is 0.190). To accomplish this, select **Tool – Options...** and set the parameters shown right before you create the dimension. Deactivate the tolerance after the dimension has been created.

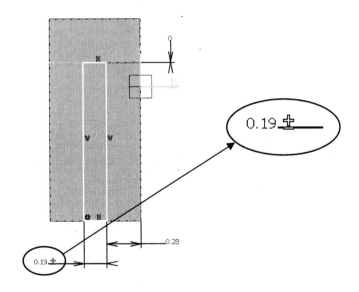

7) After completion, use **Save Management...** to save you product and parts. Your part and specification tree should look like this.

8) Enter the *Assembly Design* workbench by double clicking on **Tool Post**.

9) Unhide **Part1 (Tool Post Body)** and use the compass to move the *Wedge* away from the *Tool Post Body*.

Section 4: Part 3 (Bolt)

1) Select the **Part** icon and then click on **Tool Post** in your specification tree. You can select **No** in the *New Part: Origin Point* window.

2) Hide *Part1 (Tool Post Body)* and *Part2 (Wedge)*.

3) Rename your part **Bolt**.

4) Expand *Part3 (Bolt)* and double click on its **PartBody**. You are in the *Part Design* workbench and ready to draw the Part3.

5) Model the Bolt. Notice that the bolt shaft is toleranced. Use a basic size of 0.500 and the same procedure as before to apply the tolerance. The shaft also has threads. Use the **Thread/Tap** command to apply them.

6) After completion, use **Sa̲ve Management...** to save you product and parts. Your part and specification tree should look like this.

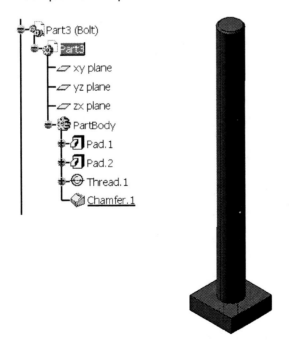

7) Enter the *Assembly Design* workbench by double clicking on **Tool Post**.

8) Unhide *Part1 (Tool Post Body)* and *Part2 (Wedge)*, and use the compass to move the *Bolt* away from the other parts.

9) Save all.

Section 5: Part4 through Part9

1) Model parts 4 through 9 and add them to the assembly as before. Note that parts 6, 7, and 8 are standard parts. CATIA contains a library of standard metric parts that can be accessed. However; this is an English drawing. Thus, we must draw the standard parts. Key steps for creating each part, the final specification tree, and final part are shown in a series of figures given below.

Part 4 (Tool Post Screw)

Part 5 (Adjusting Scew)

Part 6 (Hex Nut)

Part7 (Set Screw)

Part8 (Washer)

Part9 (Slotted Washer)

Part10 (Adjusting Nut)

5/16 – 18 UNC – 3B

R.06

MEDIUM KNURL

.03 X 45° BOTH SIDES

Ø1.06 Ø.62

.31

.62

Section 6: Duplicating parts

1) After inserting and drawing all the parts, your specification tree should look like this.

2) Double click on Tool Post to enter the *Assembly Design* workbench.

3) Use your compass to move your parts in the approximate locations shown. To do this, grab the red square on the compass and move it to the part that you want to move. If the compass lines turn green, it has attached itself to the part. Use the lines and arcs to manipulate the part. When you are done, move it back to the upper right corner. It is now ready to attach it to another part.

Click on these arcs to rotate the part.

Click on these lines to move your part in the x, y or z directions.

4) Notice that we are missing one *Tool Post Screw* and one *Set Screw*. To make copies, right click on the **Part4 (Tool Post Screw)** in the specification tree and select **Copy**. Then, right click on **Tool Post** (the root assembly) and select **Paste**. Repeat for the *Set Screw*. Name your new parts as shown in the specification tree.

Section 7: Applying assembly constraints

1) Fix the *Tool Post Body*. Select the **Fix Component** ⚓ icon and then select the *Tool Post Body*. The *Fix Component* constraint symbol will appear on the *Tool Post Body*.

2) Apply a contact constraint between surface1 and surface2 as indicated in the figure. Select the **Contact**

   Constraint 🔲 icon. If an *Assistant* window comes up, **Cancel** it. Select surface1 and rotate the assembly and select surface2 (the bottom surface of the wedge). A line with the contact

   constraint symbol 🔲 at both ends will connect the two surfaces.

3) Apply a Contact Constraint 🔲 between surface3 and surface4 (the back side of the wedge) as indicated in the figure.

4) Create an offset constraint between the surface5 and surface6. Select the **Offset**

   Constraint 🔲 icon. Select surface5 and then surface6. A *Constraint Properties* window will appear. Fill in a value of *0. 5* inch in the Offset field.

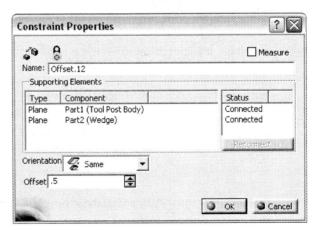

5) The part is not in position yet. Select the

Update All icon to move the part into position.

6) Apply a coincidence constraint between the *Adjusting Screw* and the counterbored hole in the *Tool Post Body*. Select the **Coincidence Constraint** icon. Select the axis of the *Adjusting Screw* (you may have to move the mouse around a bit or even zoom in before it shows up) and then the axis of the counterbored hole. A line with the coincidence constraint symbol at both ends will connect the two axes.

Adjusting Screw

Adjusting Nut

Slotted Washer

7) Apply a **Contact Constraint** between the bottom surface of the *Adjusting Screws* head (you will probably have to zoom in to be able to select it) and the bottom surface of the counterbored hole in the *Tool Post Body*.

8) Apply the appropriate **Coincidence Constraints** between the *Adjusting Screw*, and the *Slotted Washer* and *Adjusting Nut*. Remember to choose the axes and zoom in when necessary.

9) Apply the appropriate **Contact Constraints** between the *Wedge* and *Slotted Washer* and the *Slotted Washer* and the *Adjusting Nut*. Use the

   Update All to move the parts to their proper positions.

10) Use the appropriate constraints to place the two Tool Post Screws, the Bolt, the Washer, and the Hex Nut into their functional positions.

11) Apply **Coincident Constraints** to the *Set Screws* and the holes in the *Tool Post Body*.

12) Create an offset constraint between the end of the *Set Screws* and surface3. Select the

   Offset Constraint icon. Select the end of one *Set Screw* and then surface3. A *Constraint Properties* window will appear. Fill in a value of *-0.125* inch in the Offset field. Repeat for the other *Set Screw*.

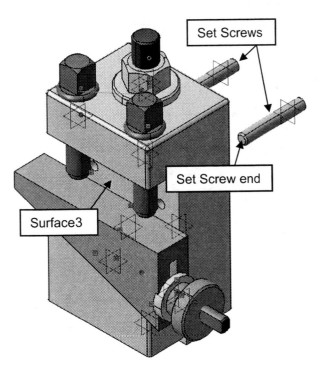

13) Hide all the constraint symbols. Right click on **Constraints** in the specification tree and select **H**ide/**Show**.

Chapter 5: ASSEMBLY DESIGN FUNDAMENTALS

Tutorial 5.2: Advanced Commands

Featured Topics & Commands

Prerequisite Knowledge & Commands

- Entering workbenches
- The *Sketcher* workbench and associated commands
- The *Part Design* workbench and associated commands
- Moving and rotating parts with the compass
- Product structure
- Assembly constraints

The Measure toolbar

The *Measure* toolbar contains commands that measure distance, angles, mass, volume, inertia, etc... of your assembly or individual geometric element. You can keep the result of your measurement as an element in your specification tree. The commands located in the *Measure* toolbar, reading left to right, are

- Measure Between: This command measures between two geometric elements.
- Measure Item: This command measures several physical parameters of an element such as area and perimeter.
- Measure Inertia: This command measures physical parameters of a part or assembly. It will calculate the mass, volume, moments of inertia, and center of gravity among other things.

The Space Analysis toolbar

The commands located in the *Space Analysis* toolbar, reading left to right, are

- Clash: This command analyzes any part interference. If there is an actual material interference between two parts, you will be warned.
- Sectioning: This command allows you to see a 2D section, and to create a 3D section of your assembly.
- Distance and Band Analysis: This command analyzes the minimum distance between two sections or between one component and all the other components.

The Annotation toolbar

Annotations are text and symbols that can be seen in the 3D view. The commands located in the *Annotations* toolbar, reading left to right, are

- Weld Feature: This command allows you to put welding symbols on your part.
- Text With Leader: This command produces a text annotation that points to a selected component or geometric feature.
- Flag Note With Leader: A flag note allows you to add hyperlinks to your document that can be used to jump to different locations. For example, you can jump to an Excel spreadsheet or to a HTML page on the internet.
- View Container Subtoolbar: These commands allow you to define the plane that will contain your annotations.

Assembly/Parts Modeled

The assembly modeled in this tutorial is used to cover more advanced commands located in the *Assembly Design* workbench. Among other things we will apply material, calculate physical properties, analyze interferences, and create a section view.

Section 1: Creating the assembly

Section 1-1: The Body

1) Enter the Assembly Design workbench.

2) Set your units to **millimeters** (**Tools – Options...**).

3) Rename your Product **Toy Car**.

4) Add a **Part** [image] to your assembly and name it **Body**.

5) Use the **Save As...** command to create a directory called **Toy Car**. Save all of your CATProducts and CATParts in this directory. The reason that we want to save the parts in a separate directory is to prevent these parts from being saved over the parts used to construct the assembly in tutorial 5-1. For example, the *Body* of the car will be, by default, saved under the name Part1.CATPart. There is already a Part1.CATPart saved from the previous tutorial.

6) Draw the following **Profile** on the **yz plane** and **Pad** it to a length of **98 mm**.

7) **Edge Fillet** the two side faces of the truck with a radius of **10 mm**.

8) Draw and **Pocket** the following sketch.

9) Enter the Sketcher on the top of the trucks bed. Project the face of the bed as a construction element and **Offset** it to the inside by **10 mm**. Turn the offset into a standard element.

10) **Pocket** the offset to a depth of **30 mm**.

11) Make the *Body* plastic. Select **PartBody** in the specification tree and select the **Apply Material** icon. In the *Library* window, select the **Other** tab and then **Bright Plastic**. Select the **Apply Material** button, then **OK**.

12) Change the color of the body to red.

Section 1-2: Wheel Assembly

1) Double click on **Toy Car** to get back into the *Assembly Design* workbench.

2) Add a new **Product** to your root assembly and name it **Wheel Assembly**. (Note: I renamed Toy Car to include Product1 just for clarity. You do not have to do this.)

3) Add a new **Part** to your *Product2 (Wheel Assembly)* and name it **Axle**.

4) Double click on the **PartBody** under *Part2 (Axle)* to enter the *Part Design* workbench.

5) Enter the **Sketcher** on the **zx plane** and sketch the axle. We will be creating a shaft that has symmetrical ends and is 150 mm long. Shown in the figure is one end.

6) **Shaft** the *Axle* and make it **Steel**.

7) Double click on **Product2 (Wheel Assembly)**. Add a new **Part** to *Product2 (Wheel Assembly)* and name it **Wheel**.

8) Double click on the **PartBody** under *Part3 (Wheel)* and sketch the following profile on the **zx plane**. **Shaft** the sketch, make the *Wheel* **Rubber**, and change its color to black.

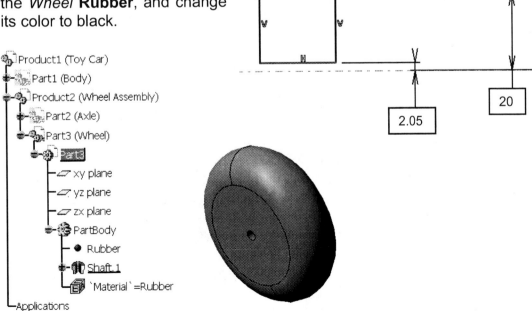

9) Double click on **Product2 (Wheel Assembly)**. Add a new **Part** to *Product2 (Wheel Assembly)* and name it **Snap Ring**.

10) Double click on the **PartBody** under *Part4 (Snap Ring)* and sketch the following profile on the **yz plane**. **Pad** the sketch to a length **0.8 mm**, and make the *Snap Ring* **Steel.**

11) Double click on **Product2 (Wheel Assembly)** and then move the parts, using the compass, to approximately the positions shown.

12)Make a copy of the *Wheel* and *Snap Ring* and move them to the positions shown.

Section 1-3: Constraints

1) Apply the appropriate constraints to the parts in the *Wheel Assembly*. See the specification tree shown for the constraints that were applied. Use an offset of 1 mm between the *Wheel* and the *Snap Ring*.

2) Make a copy of the *Wheel Assembly*. Right click on **Product2 (Wheel Assembly)** and select <u>C</u>**opy**. Then, right click on **Toy Car** and select **<u>P</u>aste**. Move the sub-assembly with the compass as shown. To move the *Wheel Assembly Copy* you need to be in the *Toy Car* product.

3) Apply the appropriate constraints between the *Wheel Assemblies* and the *Body* as shown in the specification tree given. The **Coincidence** constraints are between the *Axle* and the holes in the *Body*. The **Offset** constraints are between the inside surface of the *Wheel* and the side of the *Body* and is 1 mm.

Section 2: Measuring

1) Double click on **Toy Car** to make it the active item.

2) Select the **Measure Between** icon. In the *Measure Between* window, select **Arc center** for both selection modes. Select the ends of the *Axle*. A minimum distance of 210 mm will be calculated.

3) View the front of the car (use the *right side view* icon) and then select the **Measure Between** icon. In the *Measure Between* window, select **Edge only** for both selection modes. Select the outside surfaces of the Wheels. A minimum distance of 140 mm will be calculated.

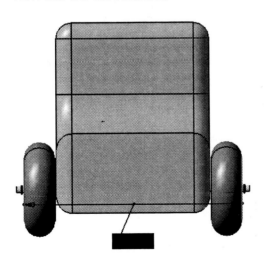

4) Select the **Measure Item** icon. In the *Measure Item* window, select the **Customize** button. In the *Measure Item Customization* window, activate the toggles shown and select **OK**. In the *Measure Item* window, select **Any geometry** as the selection mode and then select the front face of the car body. Notice that the surface area, center of gravity, and perimeter are calculated. (Note: The center of gravity may be different from what is shown. This is because we did not constrain the car body with respect to the global axis.)

5) Select **Part1 (Body)** in the specification tree and then the **Measure Inertia** icon. Notice the physical parameters that are measured. In particular notice the center of Gravity. Select **Toy Car** in the specification tree and then the **Measure Inertia** icon. Does Gz increase or decrease? Does Gy increase or decrease? Why? Why does Gx not change?

Section 3: Space Analysis

1) Make the root assembly active (Toy Car).

2) Select the **Clash** icon. In the *Check Clash* window set the type to be **Contact + Clash** and **Between all components**. Select the **Apply** button. Notice that the window changes and shows all the components that contact one another and shows no clashes.

3) Edit the car *Body* and change one of the axle holes to a diameter of **3 mm**.

After you are done, re-enter the root assembly and re-apply the **Clash** command. Notice that now there is one clash. Click on the clash and a *Preview* window will appear showing you which two parts have an interference.

4) Change the axle hole diameter back to **4.1 mm**. After you are done, re-enter the root assembly.

5) Select the **Sectioning** icon. A *Section.1* viewing window will appear and a sectioning plane will appear in the *Product* viewing window. In the *Sectioning Definition* window select the **Positioning** tab and activate the **Y** toggle. The size of the sectioning plane may be adjusted by selecting a border and moving it. The position of the sectioning plane may be changed by selecting inside the borders and moving it. Move the plane back and forth and notice that the section view changes. Change the positioning to **Z** and move the plane to see the section view change. Change the positioning to **X**. Select the coordinate axis in the middle of the sectioning plane and rotate it to see the effects (see figure on the next page). Re-select the **X** positioning and move the plane to the middle of the car. In the *Sectioning Definition* window, select the

Definition tab. Activate the **Volume Cut** icon (see figure on the next page.) Select the **Cancel** button.

Section 4: Annotations and snap shots

1) Use the compass to create an exploded view of the toy car. First move the compass to one of the wheel assemblies and move it down. Repeat for the other wheel assembly. Try to move one *Wheel* off of the *Axle*. You will find that you can not. It is part of the *Wheel Assembly*. Double click on **Product2 (Wheel Assembly)**. Use the compass to move the *Wheel* and *Snap Ring* away from the *Axel* on one side of the car. Notice that both the *Wheels* and *Snap Rings* move together. Rotate the car so that you can manipulate the other side of the *Wheel Assembly*. Double click on **Toy Car**. Select the **Product2 (Wheel Assembly)** in the specification tree and then the

   Flexible/Rigid icon. Double click on **Product2 (Wheel Assembly)**. Now move the *Wheel* and *Snap Ring* that are in the *Wheel Assembly*. Now they move independently. Repeat for the *Wheel Assembly Copy*.

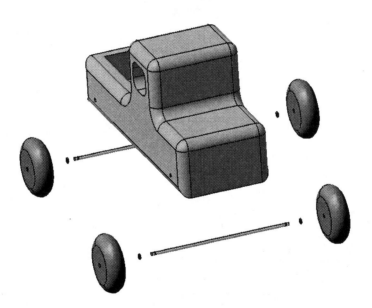

2) At the top pull down menu, select **Tools – Image – Capture**. In the Capture window, select the **Select Mode** icon. Draw a box around the exploded toy car and then select the **Capture** icon. A *Capture Preview* window will appear showing your snap shot. If this is what you want, select the **Save As** icon. In the *Save As* window, name your file and **Save**.

3) **Update All** (CTRL + U).

4) Select the **Text With Leader** icon and then select the front end of the car *Body*. In the *Text Editor* window, type **Body**. A text annotation will appear. This can be moved around and the dashed rectangle may be hidden. Right click on the annotation and select **Properties**. In the *Properties* window select the **Font** tab. Set the font size to be *10 mm*. Create another annotation for the *Wheel* as shown.

Chapter 5: Exercises

Exercise 5.1: This exercise can be performed after completing the tutorials presented in chapter 5. Model and constrain the following parts and assembly.

Exercise 5.2: This exercise can be performed after completing the tutorials presented in chapter 5. Model and constrain the following parts and assembly.

PART #	PART NAME	REQ'D	MAT
1	TROLLEY WHEEL	2	1020 STEEL
2	SIDE PLATE	2	HRS
3	CROSS HEAD	1	CRS
4	AXLE	2	CRS
5	ADJUSTING SHIM	2	HRS
6	KEEPER PLATE	2	CRS
7	BALL BEARING	2	−
8	AXLE SNAP RING	2	SPRING STEEL
9	WHEEL SNAP RING	4	SPRING STEEL
10	7/8 HEX HEAD BOLT	2	−
11	7/8 HEX HEAD NUT	2	−
12	7/8 PLAIN WASHER	8	−
13	ROUND HD MACH SCR	4	−

DRAWN BY: KIRSTIE PLANTENBERG	# OF SHEETS: 1	SHEET #: 8		SCALE: 1:2.5
PART NAME: TROLLEY	PART #: −	MAT: −		REQ'D: 1

AXEL SNAP RING
SPRING STEEL

8

TROLLEY WHEEL
1020 STEEL

1

WHEEL SNAP RING
SPRING STEEL

9

DEPARTURE
BALL BEARING
#99504

7

AXEL
CRS

4

Part#1

Ø 1.8504
1.8498

Ø 4.00

3X Ø .50
EQUALLY SPACED

1.81

.44

Ø 2.50

Ø 5.25

Ø 5.91

Ø 6.38

Ø 7.12

Part#1, Part#8 and Part#9

Part#4 and Part#7

Part#2 and Part#6

Part#3 and Part#5

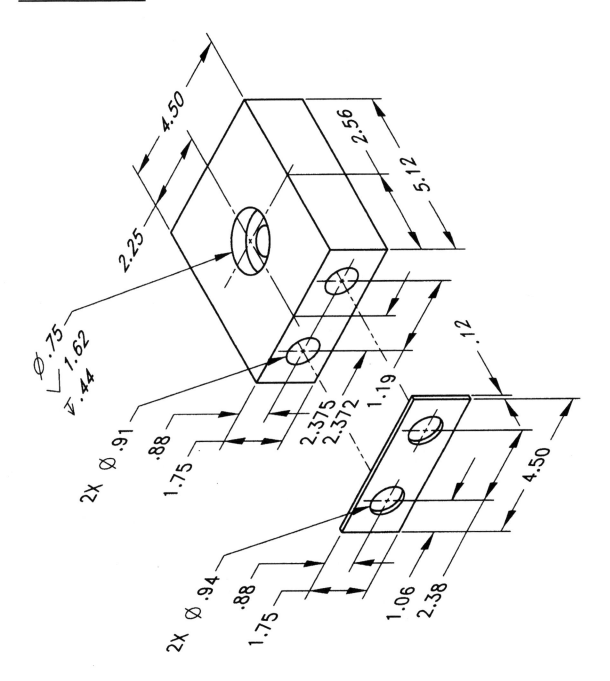

NOTES:

Chapter 6: DRAFTING FUNDAMENTALS

Introduction

Chapter 6 focuses on CATIA's *Drafting* workbench. The reader will learn how to create a simple orthographic projection and associated dimensions.

Tutorials Contained in Chapter 6

- Tutorial 6.1: Generative Drafting

NOTES:

Chapter 6: DRAFTING FUNDAMENTALS

Tutorial 6.1: Generative Drafting

Featured Topics & Commands

Prerequisite Knowledge & Commands

- Entering workbenches
- The *Sketcher* workbench and associated commands
- The *Part Design* workbench and associated commands

The Drafting Workbench

The *Drafting Workbench* allows you to create an orthographic projection or drawing (CATDrawing) directly from a 3D part (CATPart) or assembly (CATProduct). A CATDrawing contains a structure listing similar to a specification tree. The structure listing shows all the sheets and views contained in the document. CATIA enables you to create *generative views* that are associative with the 3D part, and to create *draw views* which are not associative.

The Views toolbar

The commands located in the *Views* toolbar enable you to create a variety of views and view configurations. The commands sub-toolbars in the *Views* toolbar, reading left to right, are

- Projections toolbar: The *Projections* toolbar contains commands that allow you to create different types of views. The commands, reading from left to right, are; *Front View, Unfolded View, View from 3D, Projection View, Auxiliary View, Isometric View, Advanced Front View.*
- Sections toolbar: This toolbar contains commands that allow you to create a variety of section and cut views.
- Details toolbar: The *Details* toolbar contains commands that allow you to create views that are a small portion of an existing view. The detail view is usually drawn at an increased scale.
- Clippings toolbar: The command located in the *Clippings* toolbar allow you to create removed view or removed section view. This is a small area of the part that is shown apart from the original view and is usually shown at an increased scale.
- Break view toolbar: The commands located in the *Break view* toolbar allow you to break the part in a specified location. This is usually done to save drawing space. The missing section of the part is usually uninteresting and not worth showing. There is also a command that enables you to create a broken out section.

- <u>Wizard toolbar</u>: This toolbar contains commands that allow you to select from several predefined view configurations or to define your own custom configuration. Reading from left to right, the commands are; '*View Creation Wizard*', which allows you to create custom view configurations, '*Front, Top and Left*', '*Front, Bottom and Right*', and '*All Views*'.

The Dimensioning toolbar

The commands located in the *Dimensioning* toolbar allow you to manually create dimensions and tolerances. The sub-toolbars, reading from left to right, are

- <u>Dimensions toolbar:</u> The *Dimensions* toolbar contains commands that allow you to manually create dimensions. The commands, reading from left to right, are; *Dimensions, Chained Dimensions, Cumulated Dimensions, Stacked Dimensions, Length/Distance Dimensions, Angle Dimensions, Radius Dimensions, Diameter Dimensions, Chamfer Dimensions, Thread Dimensions, Coordinate Dimensions, Hole Dimension Table,* and *Coordinate Dimension Table.*
- <u>Technological Feature Dimensions toolbar:</u> These commands allow you to dimension technological features. Technological feature dimensioning relies on the fact that technological features can specify the way they should be dimensioned. This allows you to create only realistic and customized dimensions, based on the know-how of a given field.
- <u>Dimension Editor toolbar:</u> The commands in this toolbar allow you to re-route dimensions, and create or eliminate interruptions.
- <u>Tolerancing toolbar:</u> The Tolerancing toolbar contains commands that allow you identify Datum features and to apply feature control frames with associated GD&T symbols.

Part/Drawing Modeled

The part modeled in this tutorial is used to illustrate the basic commands available in the *Drafting* workbench. You will learn how to create an orthographic projection that contains a section view, and how to create and place dimensions.

Section 1: Modeling the part

1) Model the part shown on the previous page in the *Part Design* workbench.

Section 2: Creating standard views

1) With the **PartBody** of your part selected switch to the *Drafting* workbench. In the *New Drawing Creation* window, select **Empty sheet** layout and then the **Modify...** button. In the *New Drawing* window select the **ASME** standard, an **A ANSI** format, and **Landscape** orientation. Select **OK** in both windows.

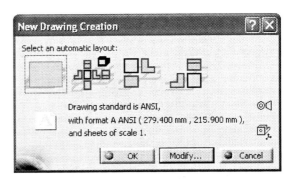

2) When the blank sheet appears, right click on **Sheet.1** in your structure listing and select **Properties**. In the Properties window make sure that the **third angle standard** is active.

3) Select the **View Creation Wizard** icon. In the *View Wizard (Step 1/2) Predefined Configuration* window, select the **Front, Top and Right** configuration icon. Select the **Next >** button. In the *View Wizard (Step 2/2): Arranging the Configuration* window, select the **Isometric view** icon. Select the **Finish** button.

4) The prompt line will read, *Select a reference plane on a 3D geometry*. At the top pull down menu, select **Window** and then your part drawing. Move your mouse over the plane that is parallel with the back of the part. An Oriented Preview will appear in the corner showing you what the front view will look like. If this is acceptable, click on the plane.

5) Your sheet should contain the front view and areas for the top, right and isometric views. These are not necessary in their final positions. The prompt line will read, *Click on the sheet to generate the view or redefine the view orientation using the arrows*. Look at the front view. If the orientation is not correct, you can use the blue circle with four arrows to reorient the front view. If the orientation is correct, click on the sheet.

6) After you accept the front view orientation by clicking on the screen, the other views will appear. The front view should be the view that is active (in a red border). If it is not active, double click on it. Move the front view towards the bottom left corner, and then move the other views as shown in the figure. To move a view, click on the view border and move the mouse. Notice that when you try to move the isometric view it is constrained by the location of the front view. Right click on the isometric view's border and select **View Positioning – Position Independently of Reference View**. Then, move the isometric view to the position shown.

7) Notice that none of the views contain hidden or center lines. Right click on the **Right view** identifier in the structure listing and select **Pr**o**perties**. In the *Properties* window, activate the **Hidden Lines, Center Lines,** and **Axis** toggles in the *Dress-up* area. Repeat for the front and top views.

Top view
Scale: 1:1

Front view
Scale: 1:1

Right view
Scale: 1:1

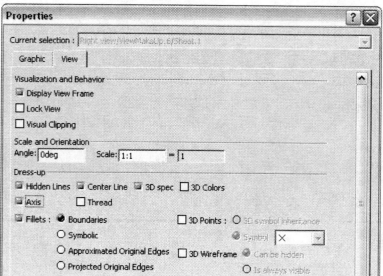

Properties

Current selection : Right view/ViewMakeUp.6/Sheet.1

Graphic | View |

Visualization and Behavior
☑ Display View Frame
☐ Lock View
☐ Visual Clipping

Scale and Orientation
Angle: 0deg   Scale: 1:1   = 1

Dress-up
☑ Hidden Lines ☑ Center Line ☑ 3D spec ☐ 3D Colors
☑ Axis   ☐ Thread
☑ Fillets : ● Boundaries   ☐ 3D Points : ○ 3D symbol inheritance
          ○ Symbolic   ● Symbol X
          ○ Approximated Original Edges ☐ 3D Wireframe ● Can be hidden
          ○ Projected Original Edges   ○ Is always visible

Section 3: Creating section views

1) Create a full section right side view. Select the **Offset Section** icon. Choose the location of your cutting plane in your front view by selecting point1 as indicated in the figure and then double clicking at point2. A shaded view will appear. Place it to the right of the right side view and click.

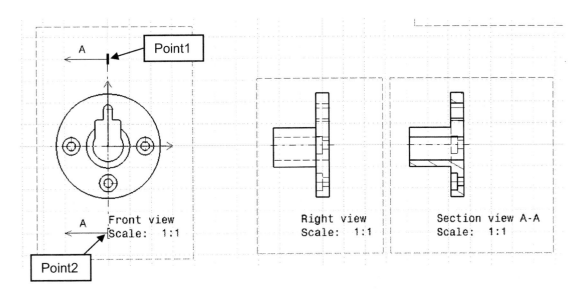

2) Change the look of the section lines. In the section view, double click on the section lines. In the *Properties* window click on the **Pattern** tab and change the angle to **45°** and the pitch to **1 mm**.

Section 4: Cleaning up the views

1) Delete the right side unsectioned view and the top view for these views are not needed. Right click on the border of the view and select **Delete**.

2) Notice that in the right side sectioned view there are two places where two lines (center and hidden) are coincident. According to drafting standards only one of those lines should be shown. In this case the hidden lines have no reason for being there (they are used to show a fillet). Enter the *Properties* window for the right side view. Deselect the **Fillet** toggle.

Hidden and centerlines overlap.

3) Section views usually do not contain hidden lines. Eliminate the hidden lines in the right side section view.

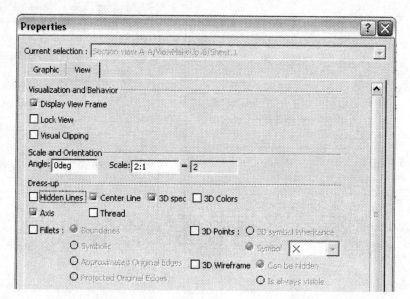

4) Change the properties of the cutting plane line. Right click on the cutting plane line and select **Properties**. In the *Properties* window, click on the **Call out** tab and set the line thickness to *3* and the line type to *5*. Set the length of the arrow base to *0.001 mm*. Set the arrow head type to solid, the length to *5 mm*, and the angle to *30°*.

5) Hide the view names. Right click on each view name and select **Hide/Show**.

6) Change the scale of the views to 2:1. Right click on Sheet.1 and select **Properties**. In the *Properties* window set the scale to *2:1*.

7) Change the scale of the views to 2:1. Right click on Sheet.1 and select **Properties**. In the *Properties* window set the scale to *2:1*.

Section 5: Dimensioning

1) Make the right side view active.

2) Draw a phantom line that indicates the full diameter of the hub. Double click on the right side section view to make it active. Draw two lines as indicated in the figure. Select both lines, right click, and select **Properties**. In the *Properties* window, set the line thickness to **1** and the line type to **5**. Commands such as line, circle, etc... are located in the *Geometry Creation* toolbar.

Phantom line indicating full diameter

3) Place the linear dimension in the right side view. Select the **Linear Dimension** ⊢⟶⊣ icon. Select the line1 and then line2. Repeat, using a similar procedure, for the 26 mm dimension. Note: You can move the arrows in and outside of the extension lines by clicking on them. You can also drag the texts outside the extension lines.

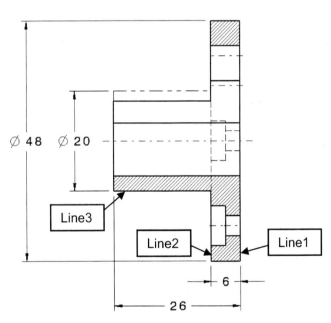

4) Select the **Diameter Dimension** ⌀⟷ icon. Select line3. The ɸ20 mm diameter dimension should appear. If it does not, select the horizontal phantom line. Repeat for the ɸ48 mm diameter dimension.

5) Activate the front view by double clicking on it.

6) Draw a **Circle** that identifies the circle of centers of the counterbored holes.

7) Right click on the circle and select **Properties**. Change the line type to **4** and the line thickness to **1**.

8) Dimension the 3 **Diameter Dimensions** ⌀ shown in the figure. Notice that the dimensions do not look like the dimension shown in the figure. Select all 3 dimensions using the Ctrl key and access the *Properties* window. Adjust the **Dimension Line** settings as shown. Position the dimensions as shown.

9) Notice that the dimension for the counterbored hole is not correctly stated. It shows only the diameter of the counterbore. Access the *Properties* window for that dimension. Select the **Value** tab and activate the **Fake Dimension** toggle. Enter a Main Value: of *4 mm*. This is the diameter of the hole. Select the **Dimension Text** tab. Enter *3X* in the top box of the Main Value field. This indicates that there are 3 counterbored holes.

10) We still need to add the counterbore diameter and depth. Select the **Text** icon. Click your mouse under the ⌀4 dimension. A *Text Editor* window will appear. At the top of the screen there should be a *Text Properties* toolbar. Click on the **Insert Symbol**

icon and select the counterbore ⊔ symbol, type a space, select the diameter symbol *⌀* , and then type *8*. Hit Enter to start a new line. Select the deep ⊽ symbol, and then type *3*.

11) Move the counterbore dimension and text down and dimension the rest of the features using the **Linear Dimension** , and **Radius Dimension** commands. Notice that the radial dimension don't look like what is shown in the figure. At the top of the screen there should be a *Dimension Properties* window. Select one of the radial dimensions and select the **Dimension Line** configuration.

12) Right click on the R1 dimension and select **Properties**. In the *Properties* window, select the **Dimension Texts** tab and enter **2X** in the left box of the Main Value field.

13) Position your views and then access the *Properties* window for each view and deactivate the **Display View Frame** toggle.

NOTES:

NOTES:

Chapter 6: Exercises

Exercise 6.1: This exercise can be performed after completing the tutorials presented in chapter 6. Create a detailed drawing of the exercises presented in chapter 2 and 3. The detailed drawing should include the appropriate views as well as an isometric view. Include section views were necessary. Dimension the orthographic projection completely.

NOTES: